Environmental Science, Engineering and Technology

Environmental Considerations in Federal Procurement

ENVIRONMENTAL SCIENCE, ENGINEERING AND TECHNOLOGY

Additional books in this series can be found on Nova's website under the Series tab.

Additional E-books in this series can be found on Nova's website under the E-books tab.

ENVIRONMENTAL SCIENCE, ENGINEERING AND TECHNOLOGY

ENVIRONMENTAL CONSIDERATIONS IN FEDERAL PROCUREMENT

JOHN G. KASPIN
EDITOR

Nova Science Publishers, Inc.
New York

Copyright © 2011 by Nova Science Publishers, Inc.

All rights reserved. No part of this book may be reproduced, stored in a retrieval system or transmitted in any form or by any means: electronic, electrostatic, magnetic, tape, mechanical photocopying, recording or otherwise without the written permission of the Publisher.

For permission to use material from this book please contact us:
Telephone 631-231-7269; Fax 631-231-8175
Web Site: http://www.novapublishers.com

NOTICE TO THE READER

The Publisher has taken reasonable care in the preparation of this book, but makes no expressed or implied warranty of any kind and assumes no responsibility for any errors or omissions. No liability is assumed for incidental or consequential damages in connection with or arising out of information contained in this book. The Publisher shall not be liable for any special, consequential, or exemplary damages resulting, in whole or in part, from the readers' use of, or reliance upon, this material. Any parts of this book based on government reports are so indicated and copyright is claimed for those parts to the extent applicable to compilations of such works.

Independent verification should be sought for any data, advice or recommendations contained in this book. In addition, no responsibility is assumed by the publisher for any injury and/or damage to persons or property arising from any methods, products, instructions, ideas or otherwise contained in this publication.

This publication is designed to provide accurate and authoritative information with regard to the subject matter covered herein. It is sold with the clear understanding that the Publisher is not engaged in rendering legal or any other professional services. If legal or any other expert assistance is required, the services of a competent person should be sought. FROM A DECLARATION OF PARTICIPANTS JOINTLY ADOPTED BY A COMMITTEE OF THE AMERICAN BAR ASSOCIATION AND A COMMITTEE OF PUBLISHERS.

Additional color graphics may be available in the e-book version of this book.

LIBRARY OF CONGRESS CATALOGING-IN-PUBLICATION DATA

Environmental considerations in Federal procurement / editor, John G. Kaspin.
 p. cm.
 Includes index.
 ISBN 978-1-61209-532-5 (hardcover)
 1. Government purchasing--Environmental aspects--United States. I. Kaspin, John G.
 JK1673.E58 2010
 352.5'3--dc22
 2010051552

Published by Nova Science Publishers, Inc. † New York

CONTENTS

Preface		vii
Chapter 1	Environmental Considerations in Federal Procurement: An Overview of the Legal Authorities and Their Implementation *Kate M. Manuel and L. Elaine Halchin*	1
Chapter 2	Green Procurement: Overview and Issues for Congress *Eric A. Fischer*	29
Chapter 3	Federal Agencies Could Improve Participation in EPA's Initiatives for Environmentally Preferable Electronic Products *John B. Stephenson*	75
Chapter 4	Statement of Casey Coleman, Chief Information Officer, U.S. General Services Administration, before the Subcommittee on Management, Organization and Procurement	87
Chapter 5	Testimony of James Jones, Deputy Assistant Administrator for Prevention, Pesticides and Toxic Substances, before the Subcommittee on Management, Organization and Procurement	91
Chapter 6	Testimony of Rick Goss, Vice President for Environment and Sustainability Information Technology Industry Council, before the Subcommittee on Management, Organization and Procurement, regarding "IT Procurement and Disposal: Application of the Federal Government's Green Policies in the Life Cycle Management of its IT Assets"	97
Chapter 7	Written Testimony of Jeff Omelchuck, EPEAT Executive Director, before the Subcommittee on Management, Organization, and Procurement	107

Chapter 8	Procuring Solar Energy: A Guide for Federal Facility Decision Makers *United States Department of Energy*	**117**

Chapter Sources **221**

Index **223**

PREFACE

Coupled with increasing concerns about the environment, the magnitude of federal spending on contracts has prompted numerous questions from Members of Congress and the public about the role of environmental considerations in federal procurement. The federal procurement system is designed to deliver on a timely basis the best value product or service to the customer, while maintaining the public's trust and fulfilling public policy objectives. This new book provides an overview of environmental consideration in federal procurement issues.

Chapter 1- Coupled with increasing concerns about the environment, the magnitude of federal spending on contracts has prompted numerous questions from Members of Congress and the public about the role of environmental considerations in federal procurement. These include to what extent do agencies consider environmental factors when procuring goods or services? What legal authorities presently require or allow agencies to take environmental factors into account when acquiring goods or services? How are existing provisions authorizing agencies to consider environmental factors implemented? This chapter provides an overview, answering these and related questions.

Chapter 2- Economic and environmental concerns have contributed to rising interest in green procurement— a term used in various ways but that may best be described as acquisition of products and services with smaller-than-average environmental footprints. Fully assessing a product or service requires integrated evaluation of cost, performance, and impacts for a set of green factors over all stages of the life cycle. Green building is an example of this approach. More generally, complexities and information gaps may constrict assessment options. However, where choices are comparative, partial assessments may often suffice. Because of such considerations, green procurement often emphasizes particular attributes, such as recycled content, energy efficiency, and waste reduction. Labeling and certification programs such as Energy Star, as well as other approaches, may be used to identify green products and services. While the use of green procurement appears to be increasing nationally, the success of programs is often not clear. Barriers to broader adoption include inadequate information among decisionmakers, lack of common implementation standards, real and perceived cost obstacles, and market and technical uncertainties.

Chapter 3- Advancing technology has led to increasing sales of new electronic devices. With this increase comes the dilemma of managing them at the end of their useful lives. If discarded with common trash, a number of environmental impacts may result, ranging from the loss of valuable resources to the potential release of toxic substances, such as lead. If recycled, they may be exported to countries with waste management systems that are less protective of human health and the environment that those of the United States.

Chapter 4- This is the statement of Casey Coleman, Chief Information Officer, U.S. General Services Administration, before the Subcommittee on Management, Organization and Procurement.

Chapter 5- The Environmental Protection Agency has taken a broad and ambitious approach to greening IT procurement in the Federal Government's purchasing practices, in the private sector in the US, and -- I'm pleased to note -- internationally, as well. As the nation's environmental agency, EPA has a leadership role in a number of programs designed to reduce environmental impacts across the full life-cycle of product manufacture, purchasing, use and disposal.

Chapter 6- Our member companies have long been leaders in sustainability: many exceed the requirements on environmental design and energy efficiency, and lead the way in product stewardship efforts. As a result, the Dow Jones Sustainability Index, the Financial Times Sustainability Index, and the Global 100 have consistently recognized numerous ITI members for their concrete environmental and sustainability achievements. In addition, *Newsweek* recently issued its 2009 Green Rankings of America's 500 largest corporations, based on their overall environmental performance, policies, and reputation. Our members occupy 4 of the top 5 positions, and 14 of the top 40 slots overall.

Chapter 7- This is a written testimony of Jeff Omelchuck, EPEAT Executive Director, before the Subcommittee on Management, Organization, and Procurement.

Chapter 8- Federal agency leaders are expressing growing interest in generating solar energy on their sites, motivated by both executive branch sustainability targets and a desire to lead by example in the transition to a clean energy economy. *Procuring Solar Energy: A Guide for Federal Facility Decision Makers* presents an overview of the solar project process in a concise, easy-tounderstand, step-by-step format. This guide is designed to help agency leaders turn their interest in solar into successful installations. Detailed information and sample documents for specific tasks are referenced with Web links or included in the appendixes. This guide concentrates on distributed solar generation and does not consider all the complexities of large, centralized solar energy generation.

Profit Zero Waste Alliance, supported by a grant from US EPA (more info at http://www.zerowaste.org/epeat/index.htm). As the process of criteria development moved forward, the stakeholders decided to formalize the system and criteria they were developing as a public technical standard working through the Institute of Electrical and Electronic Engineers, and in April 2006 they released public standard IEEE 1680. The standard contains both the environmental performance criteria for personal comp

Chapter 1

ENVIRONMENTAL CONSIDERATIONS IN FEDERAL PROCUREMENT: AN OVERVIEW OF THE LEGAL AUTHORITIES AND THEIR IMPLEMENTATION

Kate M. Manuel and L. Elaine Halchin

SUMMARY

Coupled with increasing concerns about the environment, the magnitude of federal spending on contracts has prompted numerous questions from Members of Congress and the public about the role of environmental considerations in federal procurement. These include to what extent do agencies consider environmental factors when procuring goods or services? What legal authorities presently require or allow agencies to take environmental factors into account when acquiring goods or services? How are existing provisions authorizing agencies to consider environmental factors implemented? This chapter provides an overview, answering these and related questions.

The federal procurement system is designed "to deliver on a timely basis the best value product or service to the customer, while maintaining the public's trust and fulfilling public policy objectives." Environmental objectives can generally be among the public policy objectives that factor into federal procurement. However, they are not necessarily the most significant objectives overall or in any specific procurement. There are numerous other objectives (e.g., obtaining high quality goods and services at low prices, promoting American manufacturing, protecting small businesses, fostering affirmative action) that can also factor into procurement decisions. The relationship and prioritization among these different objectives is presently unclear.

Various legal authorities currently require or allow contracting officers to take environmental considerations into account when procuring goods or, less commonly, services. These authorities can be broadly divided into three categories: (1) "attribute-focused" authorities, generally requiring agencies to avoid or acquire products based on their environmental attributes (e.g., ozone-depleting substances, recovered content); (2) general contracting authorities, allowing agencies to purchase goods with certain environmental attributes when they have bona fide requirements for such goods; and (3) responsibility-related authorities, which require agencies to avoid certain dealings with contractors that have been debarred for violations of the Clean Air or Clean Water Acts. "Attribute-focused" authorities arguably do not deprive vendors of ineligible products of due process or equal protection in violation of the U.S. Constitution. However, certain preferences for products with desired environmental attributes, or vendors of such products, could violate procurement integrity regulations and the Competition in Contracting Act if not based in statute. Use of evaluation factors based on environmental considerations is possible in negotiated procurements, but subject to certain conditions, and the reportedly lower lifecycle costs of "green" products do not, per se, mean their acquisition is justified on a "best value" basis.

Agencies generally implement existing authorities by relying on third-party designations of products with specific environmental attributes and by using standard purchasing methods, including bilateral contracts, the Federal Supply Schedules, and government-wide commercial purchase cards.

The 111th Congress has held hearings on the environmental and related effects of federal procurement practices, and Members have introduced legislation that could expand the preferences given to "sustainable" products in certain federal procurements (H.R. 5136, § 833; H.R. 5280) and assist the "green" procurement efforts of state and local governments (H.R. 1766, S. 1830). President Obama has also issued an executive order (E.O. 13514) calling for the federal government to "leverage agency acquisitions to foster markets for sustainable technologies and environmentally preferable materials," among other things.

INTRODUCTION

Coupled with increasing concerns about the environment,[1] the magnitude of federal spending on contracts[2] has prompted numerous questions from Members of Congress and the public about the role of environmental considerations in federal procurement. These include to what extent do agencies consider environmental factors when procuring goods or services? What legal authorities presently require or allow agencies to take environmental factors into account when acquiring goods or services? How are existing provisions authorizing agencies to consider environmental factors implemented? This chapter provides an overview, answering these and related questions. It does not address green building initiatives,[3] energy-savings performance contracts,[4] policy documents,[5] agency-specific laws,[6] or environmental laws of general applicability that effectively shape the products available for purchase.[7]

The 111th Congress has held hearings on the environmental and other effects of federal procurement practices,[8] and Members have introduced legislation that would allow state and local governments to use the Federal Supply Schedules to acquire "environmentally preferable 'green'" products (H.R. 1766, S. 1830); "favor" or give "preference" to "sustainable" goods in certain federal procurements (H.R. 5136, § 833; H.R. 5280); and require certain contractors operating federal facilities to comply with requirements regarding renewable energy and energy efficiency (H.R. 5280). President Obama has also issued an executive order on "Federal Leadership in Environmental, Energy, and Economic Performance," calling for the federal government to

... increase energy efficiency; measure, report, and reduce their greenhouse gas emissions from direct and indirect activities; conserve and protect water resources through efficiency, reuse, and stormwater management; eliminate waste, recycle, and prevent pollution; leverage agency acquisitions to foster markets for sustainable technologies and environmentally preferable materials, products, and services; [and] design, construct, maintain, and operate high performance sustainable buildings in sustainable locations.[9]

PLACE OF ENVIRONMENTAL CONSIDERATIONS IN FEDERAL PROCUREMENT

Fundamentally, federal procurement involves agencies acquiring[10] the goods and services they need to carry out their missions. The vision for federal acquisition, as presented in the Federal Acquisition Regulation (FAR),[11] is "to deliver on a timely basis the *best value* product or services to the customer, while maintaining the public's trust and fulfilling public policy objectives."[12] Although "best value" is not defined in this context,[13] the FAR further provides that

> All participants in the [Federal Acquisition] System are responsible for making acquisition decisions that deliver the best value product or service to the customer. Best value must be viewed from a broad perspective and is achieved by balancing the many competing interests in the System. The result is a system which works better and costs less.[14]

Environmental objectives can constitute one of the "public policy objectives" to be furthered by federal procurement, and environmental interests can be among the "competing interests" in the federal procurement system.[15] There are, however, numerous other policy objectives and interests implicated in federal procurement. These include, but are not limited to, obtaining high quality goods at low prices through competition;[16] protecting American manufacturing from foreign competition;[17] ensuring opportunities for small businesses;[18] protecting employees or prospective employees of government contractors from discrimination and promoting affirmative action;[19] ensuring that workers on certain federal construction and manufacturing contracts are adequately paid and have adequate working conditions;[20] ensuring that the government does not purchase the products of child labor;[21] and assuring that government contractors are not also doing business with regimes whose interests are adverse to those of the United States.[22]

Currently, the relationship between these various objectives and interests is not always clear from a policy or legal perspective. From a policy perspective, "balancing ... competing interests" in the procurement system, as called for in the FAR, can have unintended consequences or lead to conflicts with other procurement objectives, as the following excerpt from Steven L. Schooner's discussion of balancing "best value" with "customer satisfaction" illustrates:

> We ... have increased our emphasis upon the concept of *best value*, or what some call *value for money*. In other words, we aspire to focus upon getting the best deal—or the best bargain—for the public's money. Such an emphasis seems logical. Unfortunately, the pursuit of best value requires greater buyer resources, from market research to negotiation. Similarly,

obtaining best value may not also please the customer (for example, if the customer requires premium quality regardless of price).

... [I]n the 1990's, the U.S. procurement system increased its emphasis on obtaining *customer satisfaction* for end users. It makes sense for buyers to try to please those for whom they serve. Unfortunately, pleasing end users, especially if the end user favors specific suppliers or demands that goods be provided quickly, frequently results in less competition and higher prices, or simply embarrassing policy decisions.[23]

Much like "best value" can be difficult to balance with "customer satisfaction," environmental considerations could prove difficult to balance with other objectives or values that are part of the federal procurement system. Similarly, from a legal perspective, the existence of multiple requirements can pose questions as to which requirement takes "precedence" when the requirements cannot be implemented simultaneously. One notable example of this involves set-aside programs for small businesses. Failure to articulate the relationship between the various set-aside programs—only one of which can be used in any procurement—has resulted in extensive litigation over which set-aside program should be used when.[24] Similar questions could arise if two different products, each of which agencies "must" purchase because of its environmental attributes, could meet a requirement,[25] or if preferences for products with particular environmental attributes were seen as thwarting or thwarted by other legal requirements (e.g., Buy American, preferences for small businesses).

LEGAL AUTHORITY TO CONSIDER ENVIRONMENTAL FACTORS

Various legal authorities currently require or allow contracting officers to take environmental considerations into account when procuring goods or, less commonly, services.[26] These authorities can be broadly divided into three categories: (1) "attribute-focused" authorities, generally requiring agencies to avoid or acquire products based on their environmental attributes (e.g., ozone-depleting substances, recovered content); (2) general contracting authorities, allowing agencies to purchase goods with certain environmental attributes when they have bona fide requirements for such goods; and (3) responsibility-related authorities, which require agencies to avoid certain dealings with contractors that have been debarred from government contracting for violations of the Clean Air Act or Clean Water Act.

Attribute-Focused Authorities

Numerous statutes, regulations, and executive orders enacted or issued since the mid-1970s authorize agencies to "prefer" certain products because of their environmental attributes.[27] This generally means that agencies must purchase products with these attributes instead of competing products that lack them.[28] However, the exact nature of the preference varies by product, as discussed below and illustrated in **Table 1** and **Table A-1**. Because the attribute-focused authorities developed over time and through the actions of different branches of the federal government, they arguably do not represent a holistic framework for or ensure consistency in agencies' treatment of products or vendors on environmental grounds.

Categories of Preferred Products

A number of products are eligible for various preferences in federal procurement, discussed below, because of their environmental attributes. In some cases, the products and their attributes are defined fairly narrowly (e.g., plastic ring carriers, electric motors of 1 to 500 horsepower, solar hot water heaters),[29] although generally not so narrowly as to be identified by brand name.[30] These preferences are typically not incorporated in the Federal Acquisition Regulation (FAR), which thus serves as only a partial guide to the products preferred because of their environmental attributes. In other cases, products and attributes are defined more broadly (e.g., biobased products, recovered-content products, etc.). These broader preferences are generally incorporated in Part 23 of the FAR.[31]

Many of these preferred products have their own definitions for purposes of federal procurement, as illustrated in the Glossary below.[32] These definitions do not necessarily correspond to everyday or environmentalists' usage of these terms. Moreover, certain attributes which are currently widely discussed in environmental contexts (i.e., "green" and "sustainable") are not defined for purposes of federal procurement and are not among the attributes in terms of which current preferences are stated.[33] Some commentators have suggested that "green" products could be preferred under the existing authorities pertaining to "environmentally preferable products,"[34] which is probably the case in most circumstances. However, it is important to be clear that, absent changes in the law, any preferences given to such products are based on their being "environmentally preferable products," not "green products."

GLOSSARY[35]

Alternative fuel vehicles	Vehicles that operate solely on alternative fuel or are capable of operating either on alternative fuel or on gasoline or diesel fuel. (42 U.S.C. § 13211(3))
Alternative fuels	Methanol, denatured ethanol, and other alcohols; mixtures generally containing 85% or more, by volume, of methanol, denatured ethanol, and other alcohols with gasoline or other fuels; natural gas, including liquid fuels domestically produced from natural gas; liquefied petroleum gas; hydrogen; coal-derived liquid fuels; fuels (other than alcohol) derived from biological materials; electricity (including electricity from solar energy); or other fuel the Secretary determines is substantially not petroleum and would yield substantial energy security and environmental benefits. (42 U.S.C. § 13211(2))

Biobased products	Commercial or industrial products other than food or feed, as determined by the U.S. Department of Agriculture, that are composed, in whole or in significant part, of biological products, including renewable domestic agricultural materials (including plant, animal, and marine materials) or forestry materials. (48 C.F.R. § 2.101)
Energy Star® products	Products rated for energy efficiency under the Energy Star® program established under 42 U.S.C. § 6294a. (42 U.S.C. § 8259b(a)(2))
Energy-Efficient Products	Products[36] that meet Environmental Protection Agency (EPA) and Department of Energy (DOE) criteria for the use of the Energy Star® label or are in the upper 25 percent of efficiency for similar products as designated by FEMP. (48 C.F.R. § 2.101)
Environmentally Preferable Products	Products having a lesser or reduced effect on human health and the environment when compared with competing products that serve the same purpose. (48 C.F.R. § 2.101)
FEMP-designated products	Products designated under the Federal Energy Management Program (FEMP) of the DOE as being among the highest 25% of equivalent products for energy efficiency. (42 U.S.C. § 8259b(a)(4))
Ozone-Depleting Substances	Any substances the EPA designates in 40 C.F.R. Part 82 as Class I, including, but not limited to, chlorofluorocarbons, halons, carbon tetrachloride, and methyl chloroform; or Class II, including, but not limited to, hydrochlorofluorocarbons. (48 C.F.R. § 2.101)
Priority chemicals	Chemicals identified by the Interagency Environmental Leadership Workgroup or by an agency pursuant to Section 503 of Executive Order 13148.[37] (48 C.F.R. § 52.223-5 (a))
Recovered-content products	Products made from waste materials and by-products recovered or diverted from solid waste, not including those materials and by-products generated from and commonly used within an original manufacturing process. (48 C.F.R. § 2.101)
Toxic chemicals	Chemicals or chemical categories listed in 40 C.F.R. § 372.65. (48 C.F.R. § 52.223-5 (a))

Types of Preferences

The exact nature of the preference(s) given to products based upon their environmental attributes varies by product, but agencies could be required or encouraged to

- purchase products with the desired environmental attributes instead of competing products that lack these attributes;[38]
- avoid or minimize purchases of products with certain attributes;[39]
- draft specifications for goods or services so as to maximize the purchase and use of products with certain environmental attributes;[40]
- develop "affirmative procurement plans" to maximize the acquisition of products with certain environmental attributes;[41]
- insert clauses regarding the provision or use of designated products into certain service or construction contracts;[42]
- use certain environmental considerations as evaluation factors when considering bids or offers;[43]
- meet goals for the procurement of certain types of products;[44] or
- report agencies' performance in acquiring preferred products to executive branch authorities, Congress or congressional committees, or the public.[45]

In a few cases, agencies are also required to use contract terms that obligate contractors to certify that they have provided designated products,[46] or disclose information about certain environmental impacts of designated products.[47] **Table 1** illustrates which preferences generally apply to the major categories of preferred products included in the Glossary.

These preferences are seldom absolute, however, not even when agencies are "required" to purchase products with certain attributes. There are several reasons for this. First, the requirements themselves are generally either conditional or subject to exemptions that allow agencies to purchase products without the desired attributes in certain circumstances. Any preferences that agencies give to alternatives to ozone-depleting substances must be "cost-effective,"[48] for example, while there are exemptions allowing agencies to purchase products that do not contain biobased content if biobased products cannot be acquired competitively within a reasonable time frame or do not meet reasonable performance standards.[49] There are also exemptions for products used for certain intelligence, law enforcement, or national security purposes.[50] Second, certain preferences apply only to procurements conducted in particular places,[51] or whose price exceeds certain thresholds.[52] Third, many preferences apply only to contracts for goods. Such preferences generally apply to service contracts only when the contract involves the supply of goods or the contractor operates government-owned facilities.[53] Similarly, the requirements generally "flow down" to subcontractors only in cases involving supply contracts or contractor-operated government-owned facilities.[54]

Table A-1 provides an overview of the purchase requirements pertaining to the categories of products listed in **Table 1**, including any conditions or limitations on these requirements or product-specific exemptions thereto.

Constitutionality and Legality of Such Preferences

Such attribute-focused authorities generally would not unconstitutionally deprive vendors of competing products of due process or equal protection in violation of the U.S. Constitution. Because contractors lack property rights in prospective government contracts, they generally are not deprived of due process when the government opts to buy goods and services other than those they provide.[55] Similarly, because distinctions between vendors based on the environmental attributes of their products do not involve "suspect classifications," such as race or sex, or the exercise of fundamental rights, a court would probably not find that vendors whose products lack the desired environmental attributes are denied equal protection. Absent a suspect classification or fundamental right, a party challenging a government program on equal protection grounds must show that the program is not rationally related to a legitimate government objective by "negativ[ing] every conceivable basis which might support" the program.[56] Such challenges frequently fail because rational basis review is a deferential standard of review and "serves to invalidate only 'wholly arbitrary acts.'"[57]

Regulations or executive orders that mandate certain forms of preferential treatment for products or vendors based on environmental considerations could, however, potentially violate both procurement integrity regulations and the Competition in Contracting Act (CICA) of 1984. Subpart 3.1 of the FAR requires that "Government business shall be conducted in a manner above reproach and, *except as authorized by statute or regulation*, with complete impartiality and with preferential treatment for none."[58]

CICA is arguably even more stringent, requiring that contracts be awarded through "full and open competition" unless (1) a small business set-aside is used; (2) one of seven circumstances exist that permit other than full and open competition (e.g., sole-source, urgent and compelling need); (3). the simplified procedures for "small purchases" (generally, less than $150,000) are used; or (4) agencies use procedures "otherwise expressly authorized by *statute*."[59] Thus, while Subpart 3.1 of the FAR would permit "preferential treatment" under the authority of a regulation, CICA would generally prohibit such treatment if it resulted in other than "full and open competition." These two provisions, taken together, could effectively require that certain proposed "preferences" for products or vendors based on environmental considerations originate in statute (e.g., set-asides and, potentially, price evaluation preferences).[60] Among the major attribute-based preferences, the only one not currently based in statute is that for environmentally preferable products. However, where such products are involved, agencies must "[e]mploy acquisition strategies that ... maximize the utilization of environmentally preferable products and services (based on EPA-issued guidance)"[61] and require contractors operating or providing support services at government-owned facilities to establish "program[s] to promote cost-effective waste reduction in all operations and facilities covered by th[e] contract."[62] These types of preferences are unlikely to violate Subpart 3.1 or CICA because they do not favor certain products or vendors in the source selection process and, thereby, impermissibly restrict competition.

General Contracting Authorities

In addition to the attribute-specific authorities, there are also general contracting authorities that would allow agencies to purchase products based on environmental considerations in certain circumstances. While CICA would arguably not allow agencies to prefer certain products or vendors across the board without statutory authority,[63] it provides such explicit statutory authority for agencies to define their requirements[64] based on their needs.[65] Thus, if there were a situation where an agency required a product with specific environmental attributes, the agency could generally draft its solicitation so as to obtain that product because CICA provides them with explicit statutory authority to do so.[66] Agencies' specifications articulate their requirements to prospective contractors and form the basis upon which agencies select contractors. Only bids or offers that conform to agency specifications or statements of work are deemed "responsive" and could form the basis for the award of a government contract.

Although requirements tied to environmental attributes could potentially be used in procurements conducted by sealed bidding or negotiated procurement, there are other aspects of negotiated procurement that some commentators have suggested could be more congenial to consideration of environmental attributes.[67] While agencies using sealed bidding award contracts on the basis of price alone (i.e., to the lowest-priced qualified responsible bidder),[68] agencies conducting negotiated procurements use agency-determined evaluation factors in selecting the contractor.[69] Certain uses of evaluation factors based on environmental considerations have been upheld by the Government Accountability Office (GAO) in bid protests. In *Sunshine Kids Service Supply Company*, for example, the GAO upheld an agency's award of a contract based, in part, on consideration of the vendors' "environmental stewardship,"[70] while in *Future Solutions, Inc.*, it upheld a similar award based, in part, on consideration of the vendors' recycling programs for toners and cartridges; use of green delivery vehicles; and implementation of environmental management systems.[71] However, although agencies' use of evaluation factors tied to environmental considerations has been generally upheld,[72] agencies are subject to certain limitations in the use of such factors, the most significant of which is arguably that evaluation factors must "represent [a] key [area] of importance and emphasis ... and [s]upport meaningful comparison and discrimination between and among competing proposals."[73] In other words, any evaluation factors based on environmental considerations would have to be related to the goods or services being acquired. Additionally, agencies must also generally consider price or cost, past performance, and the quality of the product or service as evaluation factors in every procurement.[74] This means that any environmental factors would be one among many—possibly competing—factors on the basis of which the award is made.

Table 1. Legal Preferences for Major Types of Preferred Products

	Purchase Required[a]	Purchase Encouraged	Specs.	Procure. Plans[b]	Preferences Contract Clauses	Eval. Factors	Goals	Reporting	Vendor Certs.	Vendor Disclosures
Alternative fuel; alternative fuel vehicles 42 U.S.C. § 6374(a); 42 U.S.C. § 6374e; 42 U.S.C. § 7588(f); 42 U.S.C. § 13212; 42 U.S.C. § 13527(a); 42 U.S.C. § 17053; 42 U.S.C. § 17142; Exec. Order 13423	X						X	X		X
Alternatives to ozone-depleting substances 48 C.F.R. §§ 23.803—23.804; 48 C.F.R. §§ 52.223-11—52.223-12; Exec. Order 13514		X	X		X			X		X
Alternatives to toxic and priority chemicals 48 C.F.R. §§ 23.904—23.906; 48 C.F.R. §§ 52.223-13—52.223-14; Exec. Order 13514		X			X		X	X	X	X
Biobased products 7 U.S.C. §§ 8102(a) & (g); 48 C.F.R. §§ 23.404 & 406; 48 C.F.R. §§ 52.223-1—52.223-2 Exec. Order 13514	X		X	X	X		X	X	X	X
Energy Star® and energy-efficient products 42 U.S.C. §§ 17143-17144; 42 U.S.C. § 8259b(b)(3); 48 C.F.R. §§ 23.203 & 23.206; 48 C.F.R. § 52.223-16; Exec. Order 13514	X				X	X	X	X		X

Product Type	Preferences							
	Purchase Required[a]	Purchase Encouraged	Specs.	Procure. Plans[b]	Contract Clauses	Eval. Factors	Goals	Reporting	Vendor Certs.	Vendor Disclosures
Environmentally preferable products *48 C.F.R. §§ 23.703—23.706; Exec. Order 13514*		X			X		X			X
EPEAT-registered products *48 C.F.R. § 23.705; Exec. Order 13423*	X				X		X			X
Recovered-content products *42 U.S.C. § 6962(c), (d), (g) & (i); 48 C.F.R. §§ 23.404 & 23.406; 48 C.F.R. §§ 52.223-4 & 52.223-17; Exec. Order 13514*	X		X	X	X		X	X	X	

Source: Congressional Research Service, based on various sources cited in Table 1.

Notes: Table 1 includes only those preferences based in law. It does not include any additional preferences that may have been created under various government-wide or agency-specific policy documents.

a. Even when agencies are "required" to purchase products with certain attributes, these requirements are generally conditional or subject to exemptions which allow them to purchase products without these attributes in certain circumstances, as is discussed below.

b. Such plans are distinct from the "environmental management systems" that agencies are required to implement under Executive Order 13423.

Similarly, while some commentators have suggested that the focus on "best value" in negotiated procurements would result in *de facto* preferences for products with desirable environmental attributes, such commentators may confuse "best value" as the goal of all federal procurements and "best value" as a synonym for the cost/technical tradeoff process involved in negotiated procurements.[75] "Best value" is the goal of all federal procurements, but there is no special legal authority for implementing this goal independent of existing statues and regulations, of which only the attribute-focused statutes would authorize agencies to prefer products or vendors based on environmental considerations. "Best value" is also the desired result of the cost-technical tradeoff process in negotiated procurements, but the use of this process is subject to all the limitations discussed above (e.g., evaluation factors must represent a key area of importance and emphasis). This makes it unlikely that the reportedly lower life cycle costs of environmentally sound products would necessarily result in the selection of such products in all or even most procurements.[76]

Responsibility-Related Authorities

While agencies do not have authority to prefer certain contractors over others based on environmental considerations, they are required to avoid dealings with environmentally irresponsible contractors in certain circumstances. Agencies are prohibited by statute from contracting with vendors who have been debarred from federal contracts by the Administrator of the Environmental Protection Agency (EPA) for certain violations of the Clean Air and Clean Water Acts.[77] Such debarments are mandatory for specified violations; last until the EPA Administrator certifies the condition is corrected; and can be waived only if the President determines that doing so is in the "paramount interests of the United States" and notifies Congress.[78] However, these debarments apply only to the vendors' operations at the facility at which the violations occurred.[79] This means that vendors with multiple facilities are not excluded from all federal contracts.

There could also potentially be circumstances in which a particular contractor who is not debarred from federal contracting is found to be nonresponsible for purposes of the award of a federal contract because of environmental considerations. Federal law requires that agencies determine that prospective contractors are "responsible" before awarding any contract.[80] This determination is based on a number of factors, including the contractors having the necessary technical skills and facilities to perform the contract, or the ability to obtain them.[81] Certain contractors could conceivably be found nonresponsible for certain contracts because of environmental considerations under these factors. However, because responsibility determinations must be made on the basis of the most recent information available, vendors who have remedied previous environmental problems could not repeatedly be found nonresponsible on the basis of these problems.[82]

IMPLEMENTATION OF EXISTING AUTHORITIES

Implementation of the attribute-specific and general authorities that could allow agencies to prefer certain products or vendors based on environmental considerations involves two components: identification of prospective products and contractors and implementation of various purchasing methods.

Identification of Products and Contractors

In the case of the attribute-specific authorities, contracting officers generally rely on third-party designations of eligible (or ineligible) products, rather than making their own determinations of which products qualify on a case-by-case basis. In fact, the statutes and executive orders providing such authority often require both that (1) one agency, with appropriate technical expertise, designate eligible products[83] and (2) other agencies purchase these products.[84] Where recovered-content products are involved, for example, the EPA designates eligible products, while the U.S. Department of Agriculture designates biobased products. When relying on the general contracting authorities, program managers or other program personnel identify their requirements and communicate these requirements to the contracting officer, who incorporates them into a solicitation.

Parties excluded from government contracting because of violations of the Clean Air and Clean Water Acts, among other things, are listed in the Excluded Parties List System (EPLS).[85] Responsibility determinations are made on a case-by-case basis by contracting officers considering information included in the Federal Awardee Performance and Integrity Information (FAPIIS),[86] as well as information submitted by the prospective contractor and from other sources.[87]

Purchasing Methods

When purchasing products or services under either the attribute-focused or general contracting authorities, agencies rely on the same vehicles or methods generally available for their use in purchasing goods or services.[88] This includes (1) bilateral contracts; (2) the Federal Supply Schedules; and (3) government-wide commercial purchase cards. When determining which of these options to use, contracting officers consider various factors, such as the nature or type of the agency's requirements (i.e., goods or services, or both); the anticipated cost (or price); and the complexity of the procurement.

Contract

Probably the best-known procurement vehicle is the bilateral contract, which "means a mutually binding legal relationship obligating the seller to furnish the supplies or services (including construction) and the buyer to pay for them."[89] Contracts are the end result of a process that begins when agencies identify their requirements and craft solicitations to procure goods or services meeting these requirements. Solicitations identify, or describe, what agencies want to buy and also include applicable information, instructions, or guidance related to, for example, packaging and marking, inspection and acceptance, contract administration, special contract requirements, applicable contract clauses, representations and certifications, and evaluation factors for award.[90] As discussed earlier, these factors may include ones that addresses environmental considerations and attributes. For example, an agency that uses the tradeoff source selection method for a specific procurement could include environmental considerations as a non- cost, or non-price, evaluation factor provided that they have a bona fide need for goods or services with specific environmental attributes.

Federal Supply Schedules

Another option available to agencies procuring goods or services with desirable environmental attributes[91] involves the General Services Administration's (GSA's) Federal Supply Schedules.[92] A schedule is an online "catalogue" that contains goods or services offered by multiple vendors.[93] Each schedule focuses on a particular category of goods and services, and GSA has established and maintains over 40 schedules, covering things such as "advertising and integrated marketing solutions" (Schedule 541) and "professional engineering services" (Schedule 871).[94]

Federal agencies can use GSA's online shopping and ordering system, GSA Advantage!, to procure goods and services off the Schedules.[95] Several special buying programs are listed on this Web page, including "Environmental," which leads to a separate Web page ("Go Environmental with GSA Advantage!").[96] This Web page enables prospective buyers to identify the type of product or service they plan to purchase and then select one or more environmentally-based criteria or filters. Among the 14 criteria listed are biobased, Energy Star® compliant, and EPEAT.

However, GSA leaves it up to vendors to determine and identify, as applicable, the environmental attributes of the products or services they provide. GSA notes that, for some products, "vendors denote whether the product meets the specifications and determine which symbols [environmental criteria] to display."[97] Elsewhere on its website, GSA offers the following caveat regarding vendors' claims about environmental attributes:

> To assist customers' [agencies'] efforts in complying with the requirements of environmental laws and Executive Orders (considering price, availability, and performance requirements), Schedule contractors have been requested (where possible and/or feasible) to identify items that: Have recycled content (e.g., EPA-designated items with specific content requirements); Are energy and/or water saving (e.g., Energy Star); [or] Have reduced pollutants (e.g., low volatile organic compounds (VOCs) and chromate-free).
>
> *Note*: Customers should review contractor literature and contact the contractor directly to obtain complete information regarding environmental claims.[98]

Moreover, recent court decisions have held that vendors' misrepresentation of their products on GSA Advantage! is not sufficient for liability under the False Claims Act absent government purchases of the misrepresented products through the site.[99]

Governmentwide Commercial Purchase Cards

Agency personnel authorized to make "micro-purchases" can also use government purchase cards, which are similar to credit cards, to purchase so-called green products or services.[100] A micro-purchase is "an acquisition of supplies or services using simplified acquisition procedures [e.g., a purchase card], the amount of which does not exceed the micro-purchase threshold,"[101] which is generally $3,000.[102]

When using a government-wide commercial purchase card, agency personnel may "buy any commercially available supply or service not prohibited by either federal or agency-specific procurement regulations."[103] While users of these cards are required to use certain mandatory sources of goods and services (e.g., Federal Prison Industries, nonprofit agencies employing people who are blind or severely disabled)[104] and ideally should carry out market research to identify the goods or services that best satisfy the agency requirements,[105] they otherwise have substantial discretion in determining what to purchase. Personnel could thus generally select goods or services based on their environment attributes, and GSA's online training for using purchase cards includes a segment encouraging agencies to buy "green."[106] However, such actions would generally be entirely within the agencies' discretion because micropurchases are generally exempted under the attribute-focused authorities discussed previously.[107]

APPENDIX

Table A-1. Environmentally Related Purchase Requirements, Conditions and Limitations, and Exemptions for Major Types of Preferred Products

Product Type	Preference	Conditions/Limitations	Type-Specific Exemptions[a]
Alternative fuels; alternative fuel vehicles	Agencies to ensure that "the maximum number of vehicles acquired annually" for federal government use are alternative fuel vehicles and meet certain goals regarding the percentage of alternative fuel vehicles acquired annually. (42 U.S.C. § 6374(a); 42 U.S.C. § 13257(a)) Agencies shall obtain clean fuel vehicles from original equipment manufacturers "to the extent practicable." (42 U.S.C. § 7588(f)) All light duty motor vehicles or medium duty passenger vehicles must be low greenhouse gas emitting vehicles. (42 U.S.C. § 13212)		Requirements for alternative fuel vehicles generally apply only to agencies with "fleets" of 20 or more vehicles. (42 U.S.C. § 6374 note; 42 U.S.C. § 13212(a)(3))

Table A-1. (Continued)

Product Type	Preference	Conditions/Limitations	Type-Specific Exemptions[a]
	Not later than October 1, 2015, agencies shall achieve at least a 20% reduction in annual petroleum consumption and a 10% increase in annual alternative fuel consumption, as well as use plug-in hybrid vehicles where such vehicles are reasonably available at reasonable cost. (42 U.S.C. § 6374e; Exec. Order 13423) Agencies operating federal fleet refueling centers shall install at least one renewable fuel pump at each center by Jan. 1, 2010. (42 U.S.C. § 17053) Agencies may not procure alternative or synthetic fuel for mobility-related uses unless the lifecycle greenhouse gas emissions associated with production and combustion of that fuel are less than or equal to emissions from equivalent conventional fuel (42 U.S.C. § 17142)		
Alternatives to ozone-depleting substances	Agencies to "give preference" to the procurement of alternative chemicals, products and manufacturing processes that reduce overall risks to human health and the environment by	Any preference program must be cost-effective (48 C.F.R. § 23.803(a)(1))	none

Product Type	Preference	Conditions/Limitations	Type-Specific Exemptions[a]
	lessening the depletion of ozone in the upper atmosphere. (48 C.F.R. § 23.803(a))		
Alternatives to toxic and priority chemicals	Government policy is to purchase supplies and services that have been produced with minimum adverse impact on community health and the environment. (48 C.F.R. § 23.904(a))	Companies need only report toxic chemical releases "to the greatest extent practicable" to be eligible for federal contracts (48 C.F.R. § 23.904(b)) Contractor certification required only in "competitive contracts" expected to exceed $100,000 (48 C.F.R. § 23.905(a))	none
Biobased products	Agencies must give preference to those items that are composed of the highest percentage of biobased products practicable or comply with USDA regulations (7 U.S.C. § 8102(a)(2)(A)(i)(II)) 100% of purchases of USDA-designated items must contain biobased content (48 C.F.R. § 23.404(b)(1))	Agencies must prefer products to the "maximum extent practicable without jeopardizing the intended use of the product while maintaining a satisfactory level of competition at a reasonable price." Products must meet reasonable performance standards and be acquired competitively, in a cost-effective manner. (48 C.F.R. § 23.403) Preference applicable only to products whose price exceeds $10,000, per item or in the aggregate. (48 C.F.R. § 23.400(a)-(b))	Contracting officer places a written justification in the contract file[b] indicating that the item cannot be acquired (1) competitively within a reasonable time frame; (2) meeting reasonable performance standards; or (3) at a reasonable price, or the USDA provides a categorical exemption for certain items procured for specific purposes. (48 C.F.R. § 23.404(b)(1)-(2); 48 C.F.R. § 23.405(b)(1)-(2))

Table A-1. (Continued)

Product Type	Preference	Conditions/Limitations	Type-Specific Exemptions[a]
Energy Star and energy-efficient products (including Federal Energy Management Program (FEMP)-designated products)	When acquiring energy-consuming products listed in the Energy Star Program or FEMP, agencies must (1) purchase Energy Star or FEMP-designated products and (2) for products that consume power in a standby mode and are listed on FEMP's Low Standby Power Devices product listing, either (A) purchase items that meet FEMP's standby power wattage recommendation or document the reason for not purchasing such items or (B) purchase items that use no more than one than one watt in their standby power consuming mode if FEMP has listed a product without a corresponding wattage recommendation. (48 C.F.R. § 23.203(a)(1)(i)-(ii))	Preference only mandatory for acquisitions in the "United States and its outlying areas;" in acquisitions outside these areas, agencies must make their "best efforts to comply." (48 C.F.R. § 23.200(b))	Agency headd determines in writing that no Energy Star or FEMP-designated product is (1) reasonably available that meets the functional requirements of the agency or (2) is cost-effective over the life of the product taking energy cost savings into account. (48 C.F.R. § 23.204(a)-(b))[c]
Environmentally preferable products	Maximize the utilization of environmentally preferable products and services based on EPA-issued guidance (48 C.F.R. § 23.703(b)(1))	Preference for environmentally preferable products must be cost-effective. (48 C.F.R. § 23.703(a))	None
Electronic Product Environmental Assessment Tool (EPEAT)-registered products)	Meet at least 95% of their annual acquisition requirement for electronic products with EPEAT-registered electronic products unless there is no	EPEAT-registered products preferred only in contracts performed in the United States, unless the agency provides otherwise. (48 C.F.R. § 705(a))	With EPEAT-registered products, agencies may establish their own procedures for granting exceptions to the

Product Type	Preference	Conditions/Limitations	Type-Specific Exemptions[a]
	EPEAT standard for such products. (48 C.F.R. § 23.705(a))		purchase requirements, with the goal that the dollar value of exceptions granted will not exceed 5% of the total value of electronic products acquired by the agency for which EPEAT-registered products are available. (48 C.F.R. § 23.705(c))
Recovered-content products	Agencies must give preference to those items that are composed of the highest percentage of recovered material practicable consistent with maintaining a satisfactory level of competition, subject to EPA guidelines. (42 U.S.C. § 6962(c)(1)) 100% of purchases of EPA-designated items must contain recovered-material content (48 C.F.R. § 23.04(b)(1))	Agencies must prefer products to the "maximum extent practicable without jeopardizing the intended use of the product while maintaining a satisfactory level of competition at a reasonable price." Products must meet reasonable performance standards and be acquired competitively, in a cost-effective manner. (48 C.F.R. § 23.403) Preference only applicable to products whose price exceeds $10,000, per item or in the aggregate. (48 C.F.R. § 23.400(a)-(b))	Contracting officer[b] places a written justification in the contract file indicating that the item cannot be acquired (1) competitively within a reasonable time frame; (2) meeting reasonable performance standards; or (3) at a reasonable price, or the EPA provides a categorical exemption for designated items procured for certain purposes. (48 C.F.R. § 23.404(b)(1)-(2); 48 C.F.R. § 23.405(b)(1)-(2))

Source: Congressional Research Service, based on various sources cited in Table A-1.

a. For all types of products, there are exemptions for certain uses for intelligence, law enforcement, or national security purposes. See supra note 50.

b. The FAR does not specify who makes the written determination that grounds for an exemption exist, but it would appear to be the contracting officer.
c. The "required" contract clause pertaining to Energy Star and FEMP-designated energy-efficient products may indicate another possible ground for exemption in certain situations. See 48 C.F.R. § 52.223-1 5(c)(2) (allowing contractors to supply other than Energy Star and FEMP-designated products when the contracting officer approves this in writing). Required contract clauses are discussed in more detail supra note 42 and accompanying text.
d. Agency heads would appear to be able to delegate their authority to make such written determinations. This means that contracting officers may, in practice, make these determinations, as they do with biobased and recovered-content products.

End Notes

[1] *See, e.g.*, Jeffrey M. Jones, Oil Spill Alters Views on Environmental Problems, *Gallup*, May 27, 2010, *available at* http://www.gallup.com/poll/137882/Oil-Spill-Alters-Views-Environmental-Protection.aspx (reporting increases in the percentage of Americans who view environmental protection as a higher priority than energy production).

[2] Federal Contract Awards by Extent of Competition, USASpending.gov, *available at* http://www.usaspending.gov/index.php (reporting that the federal government spent $474.6 billion on contracts in FY2009).

[3] For more on this topic, see CRS Report R40 147, *Issues in Green Building and the Federal Response: An Introduction*, by Eric A. Fischer.

[4] Energy-savings performance contracts (ESPCs) are unlike other federal contracts in that they have substantially longer terms (up to 25 years) and the agency pays the contractor a percentage of the savings realized from energy- savings measures it proposes and implements for the agency. Agencies were first given special authority to contract "for the purpose of achieving energy savings" in 1986. However, even after such contracts were designated ESPCs in 1992, agencies' authority to enter into them was temporary until 2007, when it was made permanent. See Energy Independence and Security Act of 2007, P.L. 110-140, § 514, 121 Stat. 1659 (December 19, 2007) (codified at 42 U.S.C. § 8287) (permanent authority to enter ESPCs); Energy Policy Act of 1992, P.L. 102-486, § 155, 106 Stat. 2852- 55 (October 24, 1992) (designating contracts as ESPCs and extending authority to enter such contracts); Consolidated Omnibus Budget Reconciliation Act of 1985, P.L. 99-272, § 7201, 100 Stat. 142-43 (April 7, 1986) (authorizing "contracts ... for the purpose of achieving energy savings"). Presently, the Federal Acquisition Regulation (FAR) requires agencies to "make maximum use of ... energy-savings performance contracts, when life-cycle cost-effective, to reduce energy use and cost in the agency's facilities and operations." 48 C.F.R. § 23.205(a)-(c). Although the FAR attributes the requirement that agencies "make maximum use" of ESPCs to Executive Order 13123, this order was revoked in 2007, and the FAR is now the sole authority for this requirement. *See* 72 Fed. Reg. 3919, 3923 (January 26, 2007).

[5] There are numerous policy documents pertaining to environmental considerations in federal procurement. See, e.g., Department of Defense Green Procurement Program Strategy, November 2008, available at http://www.fedcenter.gov/ _kd/go.cfm?destination=ShowItem&Item_ID= 12371. However, many policy documents have no legal effect. *See, e.g.*, Chrysler Corp. v. Brown, 441 U.S. 281, 295-302 (1979) (articulating the requirements for a statement of executive branch policy to have the force of law); U.S. Dep't of Health & Human Servs. v. Fed. Labor Relations Auth., 844 F.2d 1087 (4th Cir. 1988) (holding that Office of Management and Budget Circular A-76 does not have the force of law).

[6] *See, e.g.*, 10 U.S.C. § 2922d (procurement of "nonconventional fuels" by the Department of Defense).

[7] Section 112 of the Clean Air Act, for example, requires the Environmental Protection Agency to set emission or release standards for pollutants (i.e., the National Emission Standards for Hazardous Air Pollutants (NESHAP)). However, although the General Services Administration provides a listing of NESHAP-compliant products, the purchase of such products is not a matter of procurement law in the same way that the purchase of biobased products is because these standards apply to all goods available for purchase.

[8] *See* "Cutting the Federal Government's Energy Bill: An Examination of the Sustainable Federal Government Executive Order," Subcommittee on Federal Financial Management, Government Information, Federal Services, and International Security, Senate Committee on Homeland Security and Governmental Affairs, January 27, 2010, *available at* http://hsgac.senate.gov/public/index.cfm?FuseAction=Hearings.Hearing&Hearing_id=c7cb1779-8aa1- 4250-8dfe-18e06b579af1; "IT Procurement and Disposal: Application of the Federal Government's Green Policies in the Life Cycle Management of Its IT Assets," Subcommittee on Government Management, Organization and Procurement of the House Committee on

Oversight and Government Reform, October 27, 2009, *available* at http://oversight qit-procurement-and-disposal-application-of-the-federal-governments-green-policies-in-the-life-cycle-management-of-it%20assetsq&catid=60:hearings&Itemid=28.

[9] Executive Order 13514, 74 Fed. Reg. 52117, 52117 (October 8, 2009). For more on this order, see CRS Report R40974, *Executive Order 13514: Sustainability and Greenhouse Gas Emissions Reduction,* by Richard J. Campbell and Anthony Andrews.

[10] As defined in the FAR, an "acquisition" involves "the acquiring by contract with appropriated funds of supplies or services (including construction) by and for the use of the Federal Government through purchase or lease, whether the supplies or services are already in existence or must be created, developed, demonstrated, and evaluated." 48 C.F.R. § 2.101.

[11] The FAR, which comprises Parts 1-53 of Title 48 of the *Code of Federal Regulations* (CFR), is "the primary regulation for use by all Federal Executive agencies in their acquisition of supplies and services with appropriated funds." FAR, "Foreword."

[12] 48 C.F.R. § 1.102(a) (emphasis added). Similarly, but more specifically, the FAR also provides that the federal acquisition system is intended to
(1) Satisfy the customer in terms of cost, quality, and timeliness of the delivered product or service by, for example—
  (i)   Maximizing the use of commercial products and services;
  (ii) Using contractors who have a track record of successful past performance or who demonstrate a current superior ability to perform; and
  (iii) Promoting competition;
(2) Minimize administrative operating costs;
(3) Conduct business with integrity, fairness, and openness; and
(4) Fulfill public policy objectives.
48 C.F.R. § 1.102(b).

[13] The term "best value" is also used in federal procurement statutes and the FAR to refer to source selection methods used in negotiated contracting, including lowest price technically acceptable source selection and tradeoff source selection. See 48 C.F.R. § 15.101. The possible role of environmental considerations in determining "best value," as used in this sense, is discussed in more detail later in this chapter. *See infra* notes 75-76 and accompanying text.

[14] 48 C.F.R. § 1.102-1(b). "Competing interests"—for which the FAR provides neither a definition nor an explanation—could potentially include a variety of participants and stakeholders, requirements, values, programs, and policies. Viewing the notion of "competing interests" broadly might also permit inclusion of Steven L. Schooner's nine objectives for government procurement: transparency, integrity, competition, efficiency, best value, customer satisfaction, wealth distribution, risk avoidance, and uniformity. *See* Steven L. Schooner, Desiderata: Objectives for a System of Government Contract Law, *Pub. L. & Legal Theory Working Paper* No. 37 (2002). Schooner is Co-Director of the Government Procurement Law Program, George Washington University Law School, Washington, D.C.

[15] The federal government has a long history of using its procurement system to promote socioeconomic goals and even the adoption of particular technologies. *See, e.g.,* James F. Nagle, *A History of Government Contracting* 57-58 (2d ed., 1999) (describing how the Continental Congress instructed the postmaster general to give stagecoaches "preference" over postriders in transporting the mail, in part because it hoped that promoting this nascent industry would "make the intercourse between the different parts of the Union less difficult and expensive than formerly").

[16] *See, e.g.,* the Competition in Contracting Act of 1984, P.L. 98-369, §§ 2701-2753, 98 Stat. 1175 *et seq.* (1984) (codified, as amended, at 41 U.S.C. §§ 251 *et seq.* and 10 U.S.C. §§ 2304 *et seq.*). Title 41 applies to civilian agencies; Title 10 applies to the Department of Defense and the National Aeronautics and Space Administration.

[17] *See, e.g.,* 41 U.S.C. §§ 10a-10d; CRS Report 97-765, *The Buy American Act: Requiring Government Procurements to Come from Domestic Sources,* by John R. Luckey.

[18] *See, e.g.,* The Small Business Act of 1958, P.L. 85-536, § 2(a), 72 Stat. 384 (July 18, 1958) (codified at 15 U.S.C. § 631(a)) ("[It is] the declared policy of the Congress that the Government should aid, counsel, assist, and protect, insofar as is possible, the interests of small-business concerns.").

[19] *See, e.g.,* Exec. Order 8802, 6 Fed. Reg. 3109 (June 25, 1941) (prohibiting defense contractors from discriminating on the basis of race, creed, color, or national origin); Exec. Order 11246, 30 Fed. Reg. 12319 (September 28, 1965) (requiring certain contractors to develop and implement affirmative action plans).

[20] *See, e.g.,* 41 U.S.C. §§ 35-45 (maximum working hours, safe and sanitary working conditions); 41 U.S.C. §§ 351-357 (prevailing wages).

[21] *See, e.g.,* 48 C.F.R. §§ 22.1500-22.1505.

[22] *See, e.g.,* CRS Report RS20871, *Iran Sanctions,* by Kenneth Katzman.

[23] Schooner, *supra* note 14, at 11 (emphasis in original).

[24] *See, e.g.*, CRS Report R40591, *Set-Asides for Small Businesses: Recent Developments in the Law Regarding Precedence Among the Set-Aside Programs and Set-Asides Under Indefinite-Delivery/Indefinite-Quantity Contracts*, by Kate M. Manuel.

[25] *See, e.g.*, J. Catherine Kunz, The Greening of Government Procurement, *Thomson|West Briefing Papers* No. 08-9 (August 2008), at 3, *available at* http://www.crowell.com/documents/The-Greening-of-Government-Procurement.pdf ("The types of products designated as energy-efficient by [the Federal Energy Management Program] are similar to, and often overlap with, those qualified under the Energy Star® program.")

[26] *See infra* note 53 and accompanying text for a discussion of why existing preferences seldom apply to service contracts. The type of legal authority (i.e., statute, regulation, or executive order) determines, among other things, the ease with which particular preferences may be modified or revoked. Statutory requirements can only be changed by Congress, while those in executive orders can be changed by the President at any time. *See, e.g.*, Exec. Order 12375, 47 Fed. Reg. 34105 (August 6, 1982) (revoking President Carter's preference program for fuel efficient passenger vehicles); Exec. Order 12437, 48 Fed. Reg. 36801 (August 15, 1983) (revoking President Carter's preference program for fuel efficient non-passenger vehicles).

[27] Although widely used in reference to categories of goods or services procured because of their environmental attributes, the terms "prefer" and "preference" are often undefined, or defined in specific ways for specific goods or services. *See, e.g.*, H.R. 5136, § 833 (requiring DOD to "favor" certain products based on their environmental attributes); H.R. 5280 (requiring agencies to give "preference" to "sustainable" goods in certain circumstances). When statutes do not state the form that a "preference" should take, agencies generally have "broad discretion" to craft an appropriate preference in their regulations or individual solicitations. *See, e.g.*, HAP Constr., Inc., Comp. Gen. Dec. B280044.2 (September 21, 1998) ("Where a statute requires that a preference be given to a class of potential contractors, but does not specify a particular evaluation formula, agency acquisition officials have broad discretion in selecting evaluation factors that should apply to an acquisition to effectuate the statutory mandate, and the relative importance of those factors.") (quoting US Defense Sys., Inc., Comp. Gen. Dec. B-251544; B-251938; B-251940 (March 30, 1993) (finding that the agency did not violate a statutory requirement to give U.S. contractors preference in the award of "local guard contracts" under 22 U.S.C. § 4864 when it changed the weight given to technical proficiency relative to price in evaluating offers)). "Price preference," in particular, is a term that could have various meanings depending on its context, covering everything from a presumption that prices within a certain percentage of the lowest price are not unreasonable; price differentials for products; and price evaluation adjustments based on vendors' identity. *See, e.g.*, 48 C.F.R. § 25.105 (price differentials under the Buy American Act); 15 U.S.C. § 657a(b)(3) (allowing agencies to apply a 10% price evaluation adjustment to bids or offers submitted by certified HUBZone small businesses in unrestricted competitions).

[28] Less commonly, agencies are required to avoid or minimize purchases of products with certain environmental attributes. *See infra* note 39 and accompanying text.

[29] *See, e.g.*, 7 U.S.C. § 8102(c)(1) (plastic ring carriers); 42 U.S.C. § 8259b(d)(1) (electric motors); 42 U.S.C. § 6834 (water heaters).

[30] References to brand names are arguably discouraged under federal procurement law and policy. *See, e.g.*, 48 C.F.R. § 6.302-1(c) ("[Any] acquisition that uses a brand name description or other purchase description to specify a particular brand name, product, or feature of a product, peculiar to one manufacturer does not provide for full and open competition regardless of the number of sources solicited.").

[31] The preferences for alternative fuels and alternative fuel vehicles are not, however, incorporated into the FAR.

[32] Different authorities governing the purchase of particular products sometimes contain different definitions. *Compare* Exec. Order 13514, 74 Fed. Reg. 52117, 52125-26 (October 8, 2009) ("'[A]lternative fuel vehicle' means vehicles defined by section 301 of the Energy Policy Act of 1992, as amended (42 U.S.C. 13211), and otherwise includes electric fueled vehicles, hybrid electric vehicles, plug-in hybrid electric vehicles, dedicated alternative fuel vehicles, dual fueled alternative fuel vehicles, qualified fuel cell motor vehicles, advanced lean burn technology motor vehicles, self-propelled vehicles such as bicycles and any other alternative fuel vehicles that are defined by statute) *with* 42 U.S.C. § 13211(3) ("'[A]lternative fueled vehicle' means a dedicated vehicle or a dual fueled vehicle."). The definitions in the Glossary are taken from the FAR whenever possible, but otherwise come from the statutes or executive orders under which the FAR was promulgated.

[33] For more on "green procurement," including a discussion of the various ways in which "green" could be defined for purposes of federal procurement, see CRS Report R41 197, *Green Procurement: Overview and Issues for Congress*, by Eric A. Fischer.

[34] *See, e.g.*, *id.* at 7 n.21.

[35] The Glossary includes those categories of products mentioned in Part 23 of the FAR and Executive Order 13514. Another category, consisting of "water-efficient products," appears in some sources. *See, e.g.*, 48 C.F.R. § 23.200(a)(1) ("This subpart prescribes polices and procedures for ... [a]cquiring energy- and water-efficient products and services."). The Environmental Protection Agency's (EPA's) WaterSense program labels water-efficient products. *See* WaterSense®, *available at* http://www.epa.gov/watersense. However, no legal authority articulates a preference for water-efficient products with the same level of detail that exists for the other products listed in the Glossary.

[36] When used in reference to energy-efficient products, the term "product" does not include energy-consuming products or systems designed or procured for combat or combat-related missions. *See* 42 U.S.C. § 8259b(a)(5). Executive Order 13514 effectively adopts this definition of "products" for all categories of goods preferred on environmental grounds for purposes of its requirements. *See* 74 Fed. Reg. at 52119 (excluding weapons systems from the categories of products to which purchasing preferences under Executive Order 13514 apply).

[37] Executive Order 13148 has since been revoked, but agencies continue to designate such chemicals under other authority.

[38] 48 C.F.R. § 23.404(b)(1) (requiring that 100% of purchases of USDA-designated items contain biobased content).

[39] *See, e.g.*, 48 C.F.R. § 23.803(a)(1) (requiring federal agencies to "[i]mplement cost-effective programs to minimize the procurement of materials and substances that contribute to the depletion of stratospheric ozone"); 48 C.F.R. § 11.302(a) ("Agencies must not require virgin material or supplies composed of or manufactured using virgin material unless compelled by law or regulation or unless virgin material is vital for safety or meeting performance requirements of the contract."). More commonly, however, agencies are instructed to purchase alternatives to products with undesirable attributes. *See, e.g.*, 48 C.F.R. § 23.803(b)(2) (requiring agencies to "[s]ubstitute safe alternatives to ozone- depleting substances").

[40] *See, e.g.*, 48 C.F.R. § 23.404(c) ("Agencies may use their own specifications or commercial product descriptions when procuring products containing recovered materials or biobased products. When using either, the contract should specify (1) [f]or products containing recovered materials, that the product is composed of the (i) [h]ighest percent of recovered materials practicable; or (ii) [m]inimum content standards in accordance with EPA's Recovered Materials Advisory Notices; and (2) [f]or biobased products, that the product is composed of (i) [t]he highest percentage of biobased material practicable; or (ii) USDA's recommended minimum contents standards.").

[41] *See, e.g.*, 48 C.F.R. § 23.404(a)-(e). For recovered-content and biobased products, an affirmative procurement plan must include
(i) A recovered materials and biobased products preference program;
(ii) An agency promotion program;
(iii) For Environmental Protection Agency (EPA)-designated items only, a program for requiring reasonable estimates, certification, and verification of recovered material content used in the performance of contracts. ...
(iv) Annual review and monitoring of the effectiveness of the program.
Id. at § 23.404(a)(3)(i)-(iv). The governing statutes specify that affirmative procurement plans may be either case-bycase or minimum-content based. *See, e.g.*, 7 U.S.C. § 8102(a)(2)(D)-(E) (biobased products); 42 U.S.C. § 6962(i)(1)- (3) (recovered-content products). However, at least one guidance document from the EPA describes a third type of affirmative procurement plan for recovered-content materials, one focusing on "substantially equivalent approaches" when minimum-content standards are not appropriate. *See* Comprehensive Procurement Guideline Program, October 2007, at 3, *available at* http://www.epa.gov/waste. The Proposed Policy Letter on the Acquisition of Green Products and Services issued by the George W. Bush Administration in December 2007 would have required affirmative procurement plans for all preferred types of products and re-named these plans "green purchasing plans." *See* 72 Fed. Reg. 73904, 73906 (December 28, 2007). This Policy Letter would have superseded OFPP Policy Letter 92-4, but was never finalized, in part because of opposition from industry groups. *See, e.g.*, ABA Group Says More Specific Guidance Needed in OFPP Green Procurement Plan, 89 *Fed. Cont. Rep.* 312 (March 25, 2008); Tim Kauffman, "Buy Green" Sparks Industry Backlash, *Fed. Times*, January 14, 2008, at 1. OFPP Policy Letter 92-4 remains in effect. *See* Office of Management and Budget, Policy Letter 92-4, November 2, 1994, *available at* http://www.whitehouse.gov/omb/procurement_policy

[42] *See, e.g.*, 48 C.F.R. § 23.704 ("Executive Order 13101, Section 701, requires that contracts for contractor operation of a Government-owned or -leased facility and contracts for support services at a Government-owned or -operated facility include provisions that obligate the contractor to comply with the requirements of the order."). Exec. Order 13101 has been revoked, and the FAR now constitutes the sole authority for this requirement. *See* Executive Order 13423, 72 Fed. Reg. at 3923 (revoking Executive Order 13101). Agency officials do not always insert clauses when required, and a legal doctrine—known as the "*Christian* doctrine" because of the case from which it derived—has developed to guide courts in determining whether required, but missing, clauses will be read into government contracts. Under the *Christian* doctrine, courts will read into government contracts only those clauses that represent a "deeply ingrained strand of public procurement policy." G.L. Christian v. United States, 312 F.2d 418, 426-27 (Ct. Cl. 1963) (reading the required clause allowing contracts to be terminated for the convenience of the government into a contract from which it was lacking). No court appears to have addressed the question of whether the clauses pertaining to the purchase of preferred products contained in Part 23 of the FAR are "deeply ingrained strand[s] of public procurement policy" such that they would be read into contracts from which they are lacking, although some commentators have suggested that the trend is for "required" clauses to be found such. *See, e.g.*, Stanton G. Kunzi, Losing Sight of *Christian* Values: The Evolution and (Disturbing) Implications of the *Christian* Doctrine, 1992 *Army Law.* 11 (1992).

[43] *See, e.g.*, 42 U.S.C. § 8259b(b)(3) ("The head of an agency shall incorporate ... into the factors for the evaluation of offers ... criteria for energy efficiency that are consistent with the criteria used for rating Energy Star products and for rating FEMP designated products.").

[44] *See, e.g.*, Executive Order 13514, 74 Fed. Reg. at 52119-20 (requiring agencies to "advance sustainable acquisition to ensure that 95 percent of new contract actions including task and delivery orders, for products and services with the exception of acquisition of weapon systems, are energy-efficient (Energy Star or Federal Energy Management Program (FEMP) designated), water-efficient, biobased, environmentally preferable (e.g., Electronic Product Environmental Assessment Tool (EPEAT) certified), non-ozone depleting, contain recycled content, or are non-toxic or less-toxic alternatives, where such products and services meet agency performance requirements.").

[45] *See, e.g.*, Energy Independence and Security Act of 2007, P.L. 110-140, §§ 527-528, 121 Stat. 1663-64 (December 19, 2007) (requiring annual reports to the Director of the Office of Management and Budget on agency progress in meeting requirements regarding energy-efficiency and alternative fuel, as well as reports to the House Committee on Oversight and Government Reform and the Senate Committee on Governmental Affairs).

[46] *See, e.g.*, 48 C.F.R. § 52.223-1 ("As required by the Farm Security and Rural Investment Act of 2002 and the Energy Policy Act of 2005 (7 U.S.C. 8 102(c)(3)), the offeror certifies, by signing this offer, that biobased products (within categories of products listed by the United States Department of Agriculture in 7 CFR part 2902, subpart B) to be used or delivered in the performance of the contract, other than biobased products that are not purchased by the offeror as a direct result of this contract, will comply with the applicable specifications or other contractual requirements.").

[47] 48 C.F.R. § 52.223-9 ("The Contractor, on completion of this contract, shall (1) [e]stimate the percentage of the total recovered material content for EPA-designated item(s) delivered and/or used in contract performance, including, if applicable, the percentage of post-consumer material content; and (2) [s]ubmit this estimate to._____"); 48 C.F.R. § 52.223-14(a) ("Unless otherwise exempt, the Contractor, as owner or operator of a facility used in the performance of this contract, shall file by July 1 for the prior calendar year an annual Toxic Chemical Release Inventory Form (Form R) as described in sections 313(a) and (g) of the Emergency Planning and Community Right-toKnow Act of 1986 (EPCRA) (42 U.S.C. 11023(a) and (g)), and section 6607 of the Pollution Prevention Act of 1990 (PPA) (42 U.S.C. 13106). The Contractor shall file, for each facility subject to the Form R filing and reporting requirements, the annual Form R throughout the life of the contract.").

[48] 48 C.F.R. § 23.803(a)(1).

[49] 48 C.F.R. § 23.404(b)(1)-(2); 48 C.F.R. § 23.405(b)(1)-(2). Reliance on such exemptions generally requires a written determination that certain circumstances exist. *See* 48 C.F.R. § 23.204(a)-(b) (authorizing agencies to purchase alternatives to Energy Star® and Federal Energy Management Program (FEMP)-designated products if the agency head determines in writing that no eligible products are available that meet the agency's functional requirements or are cost-effective over the life of the product taking energy cost savings into account).

[50] *See, e.g.*, Executive Order 13514, § 18(a)-(d), 74 Fed. Reg. at 52125.

[51] *See, e.g.*, 48 C.F.R. § 23.200(b) (preference for Energy Star® and FEMP-designated energy-efficient products applies only to "acquisitions in the United States and its outlying areas"). As used in the FAR, "outlying areas" include only commonwealths (i.e., Puerto Rico and the Northern Mariana Islands); territories (i.e., American Samoa, Guam, and the U.S. Virgin Islands); and minor outlying islands (e.g., Baker Island, Howard Island, Jarvis Island, etc.). 48 C.F.R. § 2.101. The "United States" generally denotes the 50 states and the District of Columbia, although it could have other meanings in specific circumstances. *Id.*

[52] *Compare* 48 C.F.R. § 23 .202 (preference Energy Star® and FEMP-designated energy-efficient products applicable to all acquisitions) *with* 48 C.F.R. § 23.400(a) (preference for USDA-designated biobased products applicable only in acquisitions where the costs of the goods exceeds $10,000 per item or in the aggregate).

[53] *See, e.g.*, 48 C.F.R. § 23.203(a)(2) (requiring contractors to supply Energy Star® and FEMP-designated energy-efficient products when agencies "contract[] for services or construction that will include the provision of energy- consuming products"); 48 C.F.R. § 23.704 (requiring contractors that operate government-owned or -leased facilities, or provide support services at such facilities, have affirmative procurement programs for environmentally preferable products). There have recently been attempts to expand these preferences to additional service contracts. *See, e.g.*, H.R. 5280; Exec. Order 13514, 74 Fed. Reg. at 52117 *et seq*. One provision of President Obama's Executive Order, in particular, could potentially be construed to reach service contracts other than those already covered under existing statutes, regulations and executive orders. 74 Fed. Reg. at 52119-20 (requiring agencies to "advance sustainable acquisition to ensure that 95 percent of new contract actions including task and delivery orders, for products and services with the exception of acquisition of weapon systems, are energy-efficient (Energy Star or Federal Energy Management Program (FEMP) designated), water-efficient, biobased, environmentally preferable (e.g., Electronic Product Environmental Assessment Tool (EPEAT) certified), non-ozone depleting, contain recycled content, or are non-toxic or less-toxic alternatives, where such products and services meet agency performance requirements."). Two other

provisions could similarly give agencies greater incentive to require or encourage contractors to use "green" products in their own operations. 74 Fed. Reg. at 52118 (allowing reductions in greenhouse gas emissions resulting from changes in contractors' manufacturing processes, utility or delivery services, modes of transportation, or supply chain activities to count toward agencies' goals for reducing such emissions); 74 Fed. Reg. at 52124 (requesting recommendations on the feasibility of using "using Federal Government purchasing preferences or other incentives for products manufactured using processes that minimize greenhouse gas emissions").

[54] Although the FAR in places suggests that such "flow down" occurs as a matter of course, it does not require agencies' prime contractors to include terms corresponding to the product preferences in their contracts with any subcontractors. *See, e.g.*, 48 C.F.R. § 52.223-15(c)("The requirements of paragraph (b) apply to the Contractor (including any subcontractor) unless ..."). Because parties to a contract can generally bind only themselves, such requirements are often necessary when the government wants certain terms to be binding on subcontractors.

[55] *See, e.g.*, Perkins v. Lukens Steel Co., 310 U.S. 113, 127 (1940) (holding that the federal government "enjoys the unrestricted power ... to determine those with whom it will deal[] and fix the terms and conditions upon which it will make needed purchases."); Chamber of Commerce of the United States of Am. v. Napolitano, 648 F. Supp. 2d 726, 736 (S.D. Md. 2009) ("[T]he decision to be a government contractor is voluntary and ... no one has a right to be a government contractor.").

[56] Lehnhausen v. Lake Shore Auto Parts Co., 410 U.S. 356, 364 (1973).

[57] Abdulah v. Comm'n of Insurance, 907 F. Supp. 13 (D. Mass. 1995). This is in contrast to "strict scrutiny," which requires the government to show that a challenged program is necessary to meet a compelling government interest. *Compare* United States v. Virginia, 518 U.S. 515 (1996) (heightened scrutiny with a classification based on sex); Dunn v. Blumstein, 405 U.S. 330, 342 (1972) (strict scrutiny when a classification based on length of residency in a state affected the exercise of the right to vote); Rothe Dev. Corp. v. Dep't of Defense, 545 F.3d 1023 (Fed. Cir. 2008) (strict scrutiny with a classification based on race) *with* Pers. Adm'r of Mass. v. Feeney, 442 U.S. 256 (1979) (rational basis review with a classification based on prior service in the military); San Antonio Indep. School Dist. v. Rodriguez, 411 U.S. 1, 29 (1973) (rational basis review with a classification based on wealth); McGowan v. Md., 366 U.S. 420, 427 (1961) (rational basis review with a classification based on geography).

[58] 48 C.F.R. § 3.101-1 (emphasis added).

[59] 10 U.S.C. § 2304(a)(1); 41 U.S.C. § 253(a)(1) (emphasis added).

[60] *See, e.g.*, Nathan S. Page, Leeding the Charge: Using Green Builder Set-Asides to Expand Sustainable Construction and Design, 39 *Pub. Cont. L. J.* 373 (2010) (advocating a set-aside program for "green builders").

[61] 48 C.F.R. § 23.703(b)(1).

[62] 48 C.F.R. § 52.223-10(b).

[63] *See supra* notes 58- 59 and accompanying text.

[64] Agencies' "requirements" are the goods or services that they need. They generally have wide discretion in defining their requirements based upon their needs, and prospective contractors generally cannot protest if an agency that needs wall-mounted writing surfaces opts to purchase or lease dry-erase boards instead of chalk boards, for example. Many attribute-focused authorities target agencies' requirements. For example, agencies that have determined they need printer paper are generally required to buy paper made with at least 30% recovered content, as opposed to paper with less or no recovered content. *See* 48 C.F.R. § 11.303(b) ("For high-speed copier paper, offset paper, forms bond, computer printout paper, carbonless paper, file folders, white wove envelopes, writing and office paper, book paper, cotton fiber paper, and cover stock, the minimum content standard must be no less than 30 percent postconsumer materials. If paper containing 30 percent postconsumer material is not reasonably available, does not meet reasonable performance requirements, or is only available at an unreasonable price, then the agency must purchase paper containing no less than 20 percent postconsumer material.").

[65] 10 U.S.C. § 2305(a)(1)(A)(iii) (agencies to "develop specifications in such a manner as is necessary to obtain full and open competition with due regard to the nature of the property or services to be acquired"); 41 U.S.C. § 253a(a)(1)(C) (same).

[66] *See, e.g.*, Crewzers Fire Crew Transport, Inc., Comp. Gen. Dec. B-402530; B-402530.2 (May 17, 2010) ("An agency has the discretion to determine its needs and the best way to meet them."); James C. Babin, Federal Source Selection Procedures in Competitive Negotiated Acquisitions, 23 *Air Force L. Rev.* 318, 326 (1982/1983) ("Agency discretion enjoys its greatest latitude perhaps in the initial phase of the creation of a source selection system or, indeed, in the initial step of any procurement. That initial step is simply the identification of the minimum requirements or needs that will satisfy the Government's desires.").

[67] *See, e.g., Green Procurement, supra* note 33, at 23.

[68] 48 C.F.R. Subpart 14.

[69] 48 C.F.R. Subpart 15.

[70] Comp. Gen. Dec., B-292141 (June 2, 2003).

[71] Comp. Gen. Dec., B-293 194 (February 11, 2004).

[72] *See also* King Constr. Co., Inc., Comp. Gen. Dec. B-298276 (July 17, 2006) ("Agency acquisition officials have broad discretion in selecting evaluation factors that will be used in an acquisition, and we will not object to the absence or presence of particular evaluation factors or an evaluation scheme so long as the factors used reasonably relate to the agency's needs in choosing a contractor that will best serve the government's interests.").

[73] 48 C.F.R. § 15.3 04(b). Agencies must also state in the solicitation whether the non-cost or non-price factors, when combined, are "(1) [s]ignificantly more important than cost or price; (2) [a]pproximately equal to cost or price; or (3) [s]ignificantly less important than cost or price." 48 C.F.R. § 15.304(e).

[74] 48 C.F.R. § 15.304(c)(1)-(3). Past performance, however, need not be considered if the contracting officer documents the reasons for omitting it. *Id.* Examples of non-cost factors used in determining the quality of the product or service include "past performance, compliance with solicitation requirements, technical excellence, management capability, personnel qualifications and prior experience." 48 C.F.R. § 15.304(c)(2).

[75] *See supra* note 13.

[76] The same is also true of the "extra points" that agencies can give vendors for exceeding the technical requirements in negotiated procurements. *See, e.g.*, The Moreland Corporation, Comp. Gen. Dec. B-283685 (December 17, 1999) (denying a protest of an award that the agency made based, in part, on extra points given to offers who exceeded a certain technical requirement). Agencies arguably can award such points to "green" products or services only when the attributes of such products exceed the technical requirements specified in the solicitation. If the solicitation included no such requirements, the "extra points" arguably could not be awarded.

[77] For more on debarment, see CRS Report RL34753, *Debarment and Suspension of Government Contractors: An Overview of the Law Including Recently Enacted and Proposed Amendments*, by Kate M. Manuel.

[78] *See* 33 U.S.C. § 368 (mandatory debarment for contractors convicted of violating 33 U.S.C. § 1319(c)); 42 U.S.C. § 7606 (mandatory debarment for contractors convicted of violating 42 U.S.C. § 74 13(c)).

[79] *Id.*

[80] 48 C.F.R. § 9.103(b) ("No purchase or award shall be made unless the contracting officer makes an affirmative determination of responsibility."). For more on responsibility determinations, see CRS Report R40633, *Responsibility Determinations Under the Federal Acquisition Regulation: Legal Standards and Procedures*, by Kate M. Manuel.

[81] 48 C.F.R. § 9.104-1(a)-(g).

[82] New Hampshire-Vermont Health Service, Comp. Gen. Dec. B-200660 (March 16, 1981). Repeated responsibility determinations based on the same "old" information could, in fact, deprive contractors of due process in violation of the U.S. Constitution. *See, e.g.*, Shermco Indus., Inc. v. Sec'y of the Air Force, 584 F. Supp. 76, 93-94 (N.D. Tex. 1984) ("[A] procuring agency cannot make successive determinations of nonresponsibility on the same basis; rather it must initiate suspension or debarment procedures at the earliest practicable moment following the first determination of nonresponsibility."); 43 Comp. Gen. 140 (August 8, 1963) (finding that multiple determinations of nonresponsibility can be tantamount to debarment).

[83] In the case of Electronic Product Environment Assessment Tool (EPEAT)-registered products, however, the government relies on designations made by a nongovernmental entity, the Green Electronics Council.

[84] *See, e.g.*, Resource Conservation and Recovery Act (RCRA) of 1976, P.L. 94-580, § 6002(e), 90 Stat. 2823 (codified, as amended at 42 U.S.C. § 6962(e)(1)) (requiring the EPA to designate eligible products). As originally enacted, RCRA required only that the EPA, in consultation with the Administrator of General Services, the Secretary of Commerce, and the Public Printer, issue guidelines "set[ting] forth recommended practices with respect to the procurement of recovered materials and items containing such materials" and provide information as to "the availability, sources of supply, and potential use of such materials and items." *Id.* A subsequent Executive Order required the EPA to issue Comprehensive Procurement Guidelines (CPGs) and Recovered Materials Advisory Notices (RMANs) as a way of providing this guidance and information. *See* Executive Order 12873, 58 Fed. Reg. 54911 (October 22, 1993). This executive order has since been revoked, but the EPA continues to issue CPGs and RMANs under other authority.

[85] 48 C.F.R. § 9.404.

[86] *See* Clean Contracting Act of 2008, P.L. 110-417, §§ 871-873, 122 Stat. 4555-58 (October 14, 2008).

[87] *See* 48 C.F.R. § 9.105-1(c). A contractor's failure to provide necessary information could result in a nonresponsibility determination because contracting officers must determine that contractors are nonresponsible when they lack information "clearly indicating that the prospective contractor is responsible." 48 C.F.R. § 9.103(b); Sec. Assistance Forces & Equip. Int'l, Inc., Comp. Gen. Dec. B-194876 (November 19, 1980).

[88] The "Acquisition" Web page found at FedCenter.gov, which is the Federal Facilities Environmental Stewardship and Compliance Assistance Center, is a potentially useful resource for agencies. It includes links to regulations, guidance, and policy, including Executive Order 13423; databases and software tools, such as a compilation of green products; directories, catalogs, and newsletters, which includes a link to GSA's environmental products Web page; Websites that purport to contain contract or procurement "language"; and

training, presentations, and briefings. The "Acquisition" Web page can be found at http://www.fedcenter.gov/programs/buygreen/.

[89] 48 C.F.R. § 2.101.

[90] Agencies are required to post solicitations for contract actions expected to exceed $25,000 on the federal government's Federal Business Opportunities (FedBizOpps) Website, at https://www.fbo.gov. 48 C.F.R. § 5.101.

[91] GSA appears to use the following terms interchangeably when referring to products with desirable environmental attributes: "environmental oriented products," "green products," and "environmentally friendly products." *See* GSA Advantage! Welcome, *available at* https://www.gsaadvantage.gov/advgsa/advantage/main /elib.do?pg=home&src=elib; *GSA Global Supply Environmental Products Catalog* v (2009). The latter publication is available through GSA's centralized mailing list service (CMLS), at http:// www.gsa.gov/Portal/gsa/ep/contentView.do?contentType= GSA_BASIC&contentId= 14842. GSA Global Supply
is a federal program administered by the General Services Administration. While Global Supply does work with commercial suppliers to develop its product line, it is a governmental program. It is not a mandatory source. Customer base: GSA Global Supply provides common-use items such as office supplies, tools, computer products, and safety equipment to federal customers, civilian and military, around the world.
GSA Global Supply FAQ, available at http://www.gsa.gov/Portal/gsa/ep/contentView.do?contentType= GSA_BASIC&contentId=28827&noc=T (emphasis in original). GSA does not define or describe what is meant by "environmental oriented products."

[92] The Federal Supply Schedules are also known as the "multiple award schedules." A multiple award schedule (MAS) "means contracts awarded by GSA or the Department of Veterans Affairs (VA) for similar or comparable supplies, or services, established with more than one supplier, at varying prices." 48 C.F.R. § 8.401.

[93] Under the Schedules program, "GSA enters into contracts with commercial firms to provide supplies (products) and services at stated prices for given periods of time. Orders are placed directly with the Schedule contractor [by the buying agency], and deliveries are made directly to the customer [buying agency]." GSA Schedules Background, *available at* http://www.gsa.gov/Portal/gsa/ep/contentView.do?contentType =GSA_BASIC&contentId=8105.

[94] Schedule List, *available at* http://www.gsaelibrary.gsa.gov/ElibMain/ scheduleList.do;jsessionid=33A61 02A323862AE57FCC6F43D225C7B.node2.

[95] GSA Advantage! is available at https://www.gsaadvantage.gov/advgsa/advantage/main/start_page.do. While the "Go Environmental with GSA Advantage!" Web page, which is mentioned above, provides buying agencies with a means for finding "environmental oriented" products or services from among the 44 supply schedules, Schedule 899 provides environmental services. Services offered through this schedule include environmental consulting services; environmental training services; materials and waste, recycling and disposal services; and remediation and reclamation services. *See* Schedule 899—Environmental Services, *available at* http://www.gsa.gov/environmentalservices. Another schedule, Schedule 03 FAC, Energy Management Services, provides, among other things, energy management training, water conservation and management services, and assistant with green building certification. *See GSA Global Supply Environmental Products Catalog* v (2009).

[96] Other "special programs," as identified by GSA and listed on the GSA! Advantage website include the American Recovery and Reinvestment Act of 2009, strategic sourcing, disaster relief, and homeland security.

[97] Go Environmental with GSA Advantage!, *available at* https://www.gsaadvantage.gov/advgsa/advantage/main/ start_page.do.

[98] Basic Schedule Ordering Guidelines, *available at* http://www.gsa.gov/schedules-ordering.

[99] United States ex rel. Crennen v. Dell Marketing L.P., 2010 U.S. Dist. LEXIS 40943 (D. Mass. April 27, 2010) (finding that the defendants did not violate the False Claims Act by misrepresenting their products' compliance with the Buy American Act on GSA Advantage! so long as the government did not purchase the products through the site); United States ex rel. Folliard v. CDW Tech. Servs., Inc., 2010 U.S. Dist. LEXIS 38233 (D.D.C., April 19, 2010) (same).

[100] 48 C.F.R. §§ 1.603-3(b) & 13.301(a).

[101] 48 C.F.R. § 2.101.

[102] *Id.* The threshold is $2,000 for construction acquisitions related to the David-Bacon Act, and $2,500 for services acquisitions subject to the Service Contract Act. Other, higher thresholds apply to acquisitions made in "support [of] a contingency operation or to facilitate defense against or recovery from nuclear, biological, chemical, or radiological attack." For any such purchase made, or any contract awarded and performed, inside the United States, the threshold is $15,000. Outside the United States, the threshold for this type of purchase is $30,000. *Id.*

[103] GSA SmartPay 2 Purchase Card, What You Can Buy, *available at* http://apps.fas.gsa.gov/we btraining/trainingdocs/ smartpaytraining/part4.cfm.

[104] GSA SmartPay 2 Purchase Card, Before Using Your Purchase Charge Card, *available at* http://apps.fas.gsa.gov/webtraining/trainingdocs/smartpaytraining/part3.cfm.

[105] 48 C.F.R. § 13.106-3(a).

[106] GSA SmartPay 2 Purchase Card, Buy Green, *available at* http://apps.fas.gsa.gov/webtraining/trainingdocs/smartpaytraining/green.cfm. Generally, this training segment encourages agencies to buy "green" products and identifies the following six components of federal "green purchasing programs": recycled content products, Energy StarTM and FEMP-designated energy efficient products, biobased products, alternative fuels and alternative fuel vehicles, environmentally preferable products and services, and non-ozone depleting substances. *Id.*

[107] *See, e.g.*, 42 U.S.C. § 6962(f) (only purchases of goods valued at over $10,000, per item or in the aggregate, are subject to the preferences under the Resource Conservation and Recovery Act).

In: Environmental Considerations in Federal Procurement
Editor: John G. Kaspin
ISBN: 978-1-61209-532-5
© 2011 Nova Science Publishers, Inc.

Chapter 2

GREEN PROCUREMENT: OVERVIEW AND ISSUES FOR CONGRESS

Eric A. Fischer

SUMMARY

Economic and environmental concerns have contributed to rising interest in green procurement— a term used in various ways but that may best be described as acquisition of products and services with smaller-than-average environmental footprints. Fully assessing a product or service requires integrated evaluation of cost, performance, and impacts for a set of green factors over all stages of the life cycle. Green building is an example of this approach. More generally, complexities and information gaps may constrict assessment options. However, where choices are comparative, partial assessments may often suffice. Because of such considerations, green procurement often emphasizes particular attributes, such as recycled content, energy efficiency, and waste reduction. Labeling and certification programs such as Energy Star, as well as other approaches, may be used to identify green products and services. While the use of green procurement appears to be increasing nationally, the success of programs is often not clear. Barriers to broader adoption include inadequate information among decisionmakers, lack of common implementation standards, real and perceived cost obstacles, and market and technical uncertainties.

As a major consumer of goods and services with significant potential impacts on the environment, including human health, the federal government could arguably influence the adoption of green procurement generally and the market for green products and services. Federal green-procurement efforts focus largely on acquisition of products, even though services account for about half of federal procurement spending. Various statutes, regulations, executive orders, and policy documents require or encourage the purchase of several types of products because of their environmental attributes. Agencies are required to purchase alternative fuels and alternative- fuel vehicles, and products that are biobased, Energy Star and energy-efficient, EPEAT (a green technology labeling program), and that contain recycled content, but acquisitions may be exempt in specified circumstances. Agencies must consider purchasing alternatives to toxic and priority chemicals and ozone-depleting substances, and environmentally preferable (EPP) products and services. Only EPP and EPEAT attempt to provide an integrative approach, rather than addressing only one or a few attributes, but they lack a specific basis in enacted statutes.

The Office of Management and Budget (OMB) provides broad guidance through various policy documents, as does the Office of the Federal Environmental Executive (OFEE), housed at the Environmental Protection Agency (EPA). For some kinds of attributes, procurement criteria are set by specific agencies. EPA, the General Services Administration (GSA), OFEE, and other agencies have databases that help identify green products. OMB requires agencies to have green procurement plans and to report annually on their activities. Those reporting requirements appear to be largely qualitative, but quantitative reports are available for recycled-content and alternative-fuels products.

Green procurement raises several policy questions, especially for federal acquisitions: (1) What are the most useful and appropriate policy goals for green procurement? (2) Are the legal authorities and other means by which different green product and service initiatives have been established the most appropriate for meeting policy goals? (3) How effectively is agency implementation and performance of green procurement being assessed? (4) How successful are current programs and initiatives at meeting policy goals? (5) Are policies on the acquisition of green services sufficient? (6) Are the policies and the methods of implementing them sufficiently harmonized and integrated? (7) Are there significant gaps in the preferences for green products and services? (8) Are there implementation methods not currently used by the federal government that should be considered? (9) Is training of the acquisition workforce sufficient?

Increasing concerns about global warming, energy independence, and public health have contributed to a rise in public interest in so-called "green" products and services, and the development of a "green" economy.[1] A number of federal programs and activities are relevant to that interest. In addition to well-known programs aimed at the public such as Energy Star (see text box below), the federal government is itself a major consumer of goods and services, spending more than $500 billion per year on procurement (**Table 1**), with annual spending increasing significantly in recent years.[2] Substantial proportions of those expenditures are on goods and services with significant environmental impacts. For example, in both FY2008 and FY2009, more than half of the funds spent on manufactured goods was for transportation equipment, and more than 15% was for information technology and other electronics and electrical products. Construction of buildings and other structures comprised 11% of spending in FY2008 and 7% in FY2009.[3]

Those levels of expenditure imply that greater use of green procurement practices could significantly reduce the environmental impacts of federal activities. Furthermore, federal policies and practices relating to green procurement may have significant impact on the adoption of such practices by state and local governments and the private sector. They might also serve to stimulate growth and innovation in green sectors of the national economy, potentially influencing the availability of green products and services in the marketplace and the development of associated industries.

Several relevant federal procurement requirements and programs have been established, beginning with a requirement enacted in the 1970s for federal agencies to purchase products with recycled content. That requirement and some additional initiatives focus largely or entirely on federal procurement, whereas others, such as Energy Star, a joint program of the

Environmental Protection Agency and the Department of Energy, were specifically designed to have a broader market focus. These two kinds of federal efforts, which may be called agency-focused and market-focused, respectively, are distinct but not completely independent. For example, Energy Star originated as a market-focused program aimed largely at consumers, but it is now included in federal regulations relating to acquisition of energy-consuming appliances and even buildings.[4]

Most of the federal initiatives focus on specific inputs or effects that a product might have, such as energy use, recycled content, or toxicity. This approach can allow for fairly straightforward implementation of a given initiative, but it may also be criticized as not paying sufficient attention to interactions or trade-offs among various factors. In addition, evaluating the effectiveness of initiatives in implementing federal green policy can be challenging.

Several factors contribute to this situation. A fundamental problem is that there is no general agreement on what green procurement is or how best to implement it. Also, several barriers exist to broader adoption, both within the federal government and in the broader economy. Those include inadequate information, lack of common standards, concerns about costs, and both market and technical uncertainties. Debate about the desirability or feasibility of meeting green policy goals such as reducing greenhouse-gas emissions may also create uncertainties.

However, this chapter does not specifically address controversies about the desirability of green policy goals, either those Congress has established in various statutes, or those specifically addressed in executive orders issued by Presidents Clinton, George W. Bush, and Obama.[5] Those controversies are complex, often longstanding, and beyond the scope of this chapter, but some of the major ones are discussed in other CRS documents.[6] This chapter also does not address issues specific to the complex area of federal procurement, which is also discussed in other CRS documents.[7]

Rather, the report focuses on procurement from a green policy perspective. It discusses the problem of what green procurement means, selected aspects of applicability to and implementation by federal agencies (both initiatives aimed largely at federal agencies and those with a broader focus), and issues that Congress may decide to address with respect to federal green procurement programs. Those issues include what green procurement means in the federal context; how programs and requirements should be established, assessed, and harmonized; and what gaps exist at present with respect to focus, implementation, and evaluation.

ENERGY STAR

Energy Star is a voluntary labeling program established by EPA in 1992. It is now a joint EPA/DOE program. It is a public/private partnership designed to overcome market barriers to the adoption of energy-efficient products and services.

Residential: The agencies work with manufacturers to identify appliances and other products that are cost effective and energy efficient. Products meeting the criteria receive an Energy Star label. The agencies provide information directly to consumers about the thousands of labeled products. Among the product categories included are office equipment, home electronics, heating and cooling (HVAC), appliances, lighting, and windows. The program has also partnered with builders to create Energy Star-qualified homes and with lenders to encourage the use of "green mortgages" to promote energy-efficient housing.

Commercial: EPA offers partnerships to businesses and other organizations that make top-level managerial commitments to adopt superior energy management. Partners continually assess energy use within their organizations and use an integrated approach in upgrading buildings. EPA provides standardized measurement tools and a recognition program to assist and promote these efforts.

Federal: The Energy Policy Act of 2005 (P.L. 109-58, known as EPACT 2005) requires federal agencies to purchase either Energy Star products or those designated as energy efficient by FEMP. The Energy Independence and Security Act of 2007 (P.L. 110-140, known as EISA) requires additionally that federal agencies lease only facilities with a recent Energy Star label.

Source: EPA, "ENERGY STAR – The Power to Protect the Environment through Energy Efficiency," July 2003, http://www.energystar.gov/ia/partners/downloads/energy

WHAT IS GREEN PROCUREMENT?

Green procurement may appear to be a somewhat fuzzy concept, given the variety of ways in which it and related terms are used. There are many different practices and criteria that are called "green," and several alternative terms that are used at least somewhat interchangeably in public discourse. In fact, the term *green* came into widespread use only in the last several years. Other terms used in various documents both historically and currently include "environmentally preferable," "sustainable," "affirmative," and "socially responsible" procurement. Those and other terms may have specific meanings in some contexts but they may also be used somewhat interchangeably in other contexts. In addition, green procurement may be used in some cases to refer to purchasing of products and services with specific attributes such as energy or water efficiency, or use of biobased or recycled materials or nontoxic chemicals, without necessarily taking other factors into account.

Such variations in usage and meaning can create significant difficulties in understanding what green procurement is and in the development and implementation of policy goals. Therefore, this section of the report develops a general conceptual framework for consideration of policy issues associated with green procurement. It draws from both public- and private-sector initiatives and analyses. It discusses examples from federal programs, including those that have a focus beyond procurement by federal agencies, as well as examples involving states and the private sector. Green-procurement activities by federal agencies are discussed specifically later, in the section on "Federal Green Purchasing Initiatives."

Table 1. Federal Procurement Spending for FY2008 by North American Industry Classification System (NAICS) Group ($ Billions)

Group	Amount FY2008	FY2009
Products	**226.9**	**217.9**
Transportation equipment	113.6	104.5
Computer and electrical equipment	33.2	33.1
Wholesale and Retail Trade	28.5	30.0
Metals	12.9	12.6
Petroleum and coal products	11.4	13.1
Machinery	6.7	6.4
Chemicals	4.9	8.3
Other	15.8	10.0
Services	**240.4**	**263.8**
Professional, Scientific, and Technical Services	141.3	150.8
Administrative and Support Services	36.8	39.3
Information	10.4	11.9
Transportation	10.3	10.9
Finance and Insurance	9.7	12.3
Real Estate, Rental, and Leasing	5.3	4.7
Health Care and Social Assistance	5.2	6.8
Waste Management and Remediation Services	5.1	9.0
Educational Services	4.5	5.2
Repair and Maintenance	4.0	4.4
Utilities	2.2	2.8
Other	5.9	5.7
Construction	**57.2**	**38.7**
Building construction	41.5	25.6
Heavy and Other Construction	15.7	13.2
Other	**13.6**	**15.9**
Total	*538.1*	*536.4*

Source: Analysis by CRS of data from Federal Procurement Data System (FPDS-NG), https://www.fpds.gov. FY2008 data are from October 2009. FY2009 data are from March 2010.

Note: Data in the FPDS-NG database are not static and may vary slightly with different runs. Figures, which are obligated dollars, should therefore be considered approximate. For descriptions of the industry groups, see U.S. Census Bureau, "2007 NAICS," September 3, 2008, http://www.census.gov/cgi-bin/sssd/naics/naicsrch?chart= 2007. Groupings in the table are derived by CRS from descriptions of sectors and subsectors in that source, supplemented by U.S. Census Bureau, "2002 Economic Census: Summary Statistics by 2002 NAICS - United States," November 7, 2005, http://www.census.gov/econ/census02/data/us/US000.HTM. All Product categories except Wholesale and Retail Trade are derived from aggregation of subsector data. All Services categories correspond to sectors except Administrative and Support Services, Waste Management and Remediation Services, Repair and Maintenance, and Other Services. The two Construction categories are derived from aggregations of subsector data within the Construction sector.

Green as a Relative Concept

The apparent fuzziness of green procurement as a concept can create uncertainties and even confusion that may make addressing policy issues difficult. To help avoid such problems, the following characterizations are used in this chapter:

- A *green product or service* can usefully be thought of as one that has a significantly smaller environmental footprint than the average or standard product or service of the same type (see **Figure 1**).
- The *environmental footprint* of a product or service can be described as its overall impact on the environment, including use of resources such as energy and water, and effects on health and other aspects of the human environment.
- Green *procurement* can be thought of as the range of processes by which such green products and services can be acquired.[8]
- Green *practices* also include related procedures and actions that result in a reduced environmental footprint, such as using products in ways that optimize their green characteristics.[9]

In this view, the industry average or *standard* at any given time is generally not considered green.[10] Consequently, as products, services, and practices currently thought of as green become standard practice within an industry, the threshold for something to be considered green would change—moving to a level of environmental impact that is lower than the new standard.[11]

Most products and services acquired through procurement activities are likely to be subject to some environmental constraints—for example, they may be subject to environmental regulations, or the industry standard may incorporate environmental protections for other reasons. Consequently, those exhibiting more *extreme impacts* are unlikely to be considered standard practice, as illustrated in **Figure 1**. At the same time, having lower impact than what is currently standard does not necessarily mean that a practice is a *sustainable* one—that is, maintaining an environmental footprint that is small enough that it will not impede future human activity and the functioning of ecosystems.[12]

Green procurement practices may also include those that have no net environmental impact (or *zero impact*, as shown in the figure), which is unattainable in most instances at present, except in some cases for some attributes such as energy (see also the section on "Life Cycle Analysis" below). For example, the Energy Independence and Security Act of 2007 (EISA, P.L. 110-140) set a goal of zero net energy use by 2030 for new federal buildings.[13] In theory, at least, a product or service could even have a positive environmental impact, at least for some elements, such as a building that produces more energy than it uses, for example via solar panels, and returns the surplus to the grid.

The relative approach to green procurement described here may appear to create difficulties for implementation to the extent it makes green a moving target, depending on the current industry average. However, it does not preclude an approach incorporating a fixed target such as zero impact.[14] It can also be seen as an application of the well-established business practice of continual improvement.[15] And it is consistent with a common feature in the development and maintenance of voluntary consensus standards,[16] which are usually reviewed and updated on a periodic basis as technology, practices, and requirements evolve. Current federal law encourages the use of such standards by federal agencies in general,[17] and some specific standards are referenced in some provisions in federal laws relating to environmental policy. [18] States may also reference such standards, for example in building codes and procurement requirements for agencies.[19] Some sets of consensus standards have been developed specifically with respect to green procurement, for example in information technology (IT), as discussed later in this chapter.

This use of green as a relative concept has several potential advantages. It allows for broad application of practices and standards to different products and services, and for adaptation to the specific conditions faced by different sectors and industries. It arguably facilitates the use of performance standards rather than design standards, and the former are generally regarded as preferable.[20] It also permits gradual and continual reduction in environmental impact over time—for example, gradual reduction in energy use or production of pollutants. Finally, it avoids the inherent problems with defining what *green* is in a more fixed or absolute sense.

However, some observers may prefer a different meaning. Some may argue that simply taking environmental issues into account is sufficient to practice green procurement. Others may express concerns about a relative approach being more difficult to apply to specific cases, presenting unpredictably shifting goals, not accounting for inherent limitations in reducing environmental footprints, or permitting slower progress toward goals than some may desire.

For example, use of recycled content in paper products would generally be regarded as a green practice. If the goal for that practice is set at a particular percentage, such as 50%, of such content, then it is relatively easy for a manufacturer to conform and for buyers to assess whether a given product does so. But if the goal is to have more recycled content than average, a manufacturer or buyer would have to know what that average was to assess what a green percentage would be. Furthermore, if the average were 75% recycled content, then only paper with higher percentages would be considered green. If the average were 100%, no paper would be considered green with respect to recycled content. Finally, setting a high specific target, such as 100%, may result in more rapid increases in average recycled content than setting a relative goal.

Such objections can be met in several ways. For example, documentary standards can be used, where targets are set by an independent entity and can be revised as conformance improves. A proxy or indicator such as a green label can be used to indicate whether a product meets the standard. That is one approach used, for example in green IT, as discussed later.

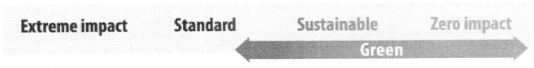

Source: CRS.
Notes: The figure shows the position of green procurement along a continuum, from those with extremely high impacts to those with none. It illustrates that green procurement denotes a range of approaches that have lower environmental impact in general than standard ones.

Figure 1. Green Procurement Relative to Other Approaches.

Also, green practices are not limited to only one factor or attribute, such as percentage of recycled materials, but involve integration of several across the life cycle of a product or service. Even if a maximum is reached for one element, options for improvement are likely to remain for others. In addition, the relative sense in which the term green is used in this chapter captures the range of meanings given to it by different stakeholders in a way that is comparatively clear and consistent and retains the utility of the concept.

Green Procurement and Environmentally Preferable Purchasing

Finally, some may argue that green procurement as described here is simply a synonym for environmentally preferable procurement, or EPP, which was introduced in the 1 990s (see "Federal Green Purchasing Initiatives" below).[21] However, EPP refers to various specific programs and initiatives,[22] usually involving explicit comparison of alternative products or services, whereas green procurement is used here in a broader sense that includes a variety of other approaches, such as zero-impact initiatives.[23] Definitions of EPP used by various nonfederal entities often follow closely the federal description, but some entities define it differently.[24]

A salient feature of the concept of green procurement described above is that it is multidimensional, usually involving an *integrated evaluation* of criteria such as cost, performance, and impact of a product or service with respect to a set of *green factors* over the entire *life cycle*, from design through disposal. These three dimensions—life cycle, green factors, and evaluation criteria—are discussed below, followed by a discussion of considerations relating to implementation.

Life Cycle Stages

The stages of the product or service life cycle can also be divided in various ways, but for purposes of this chapter, the following six stages will be discussed:

- research and design,
- raw materials acquisition,
- manufacturing,
- distribution,
- operation or use, and
- disposal or end-of-life.

The relative importance of those stages will vary depending on the kind of product or service being considered. Raw materials acquisition, manufacturing, and disposal are unlikely to be direct factors in acquisition of services in many cases, but they will play an indirect role, depending on the degree to which the service provider engages in green product-procurement practices.

For example, some consultants may use only green IT products, such as those listed under the Energy Star or EPEAT[25] labeling programs, and may follow green practices in end-of-life management of products they use. Some cleaning services may use environmentally preferable cleaning products.[26] In the latter case, failure to use green products could also have direct consequences for the service recipient, for example if the products contained chemicals with adverse health effects.

Research and Design

All products and services that come to market are designed in some manner before they are produced, and that preliminary stage may also involve performing various types of research and development. New kinds of products may require a substantial research effort before they can be designed. Such activities may involve processes with significant environmental impacts. Also, the characteristics set at the design stage can affect impacts and other criteria throughout the life cycle. As with many features, green characteristics work best if they are built in rather than added on.[27]

Nevertheless, this stage is often omitted in discussions of life cycle aspects of green procurement. That may be because in many cases, acquisitions are of products or services already available in the marketplace. However, institutional green procurement programs can have an influence at this stage, most obviously for acquisitions that require custom design and manufacture. For ready-made products and services, assessment may be made of the degree to which candidates for procurement were specifically designed both to reduce environmental impacts throughout the life cycle (sometimes called design for the environment)[28] and to facilitate recycling and other aspects of end-of-life management (sometimes called design for recycling).

Raw Materials Acquisition

The kinds of materials from which a product is manufactured may substantially affect the environmental footprint of the product. For example, mining or logging may have low to high impact, depending on the practices of the companies involved, through energy and water use, waste production, and effects on health and ecosystems. Green guidelines and practices have been developed to reduce such impacts by changing the processes by and sources from which raw materials are obtained, including increasing the use of recycled and renewable resources. Examples are requirements and programs to reduce by-catch in commercial fishing,[29] reduce mining pollution,[30] and increase use of recycled content and biobased materials in products.

Manufacturing

The impacts of processes used in manufacturing products and developing services also vary. Manufacturing processes may differ in the efficiency of input use, the amount and kind of waste generated, and environmental effects on ecosystems and human health. Such impacts may be reduced by manufacturers through various means, ranging from improvements focusing on individual factors such as amounts and sources of energy used, to integrated approaches such as lean manufacturing techniques, which aim to reduce waste and improve efficiency throughout the manufacturing process.[31]

Distribution

This life cycle stage includes the various processes involved in distributing a product or service to users once it has been manufactured or developed. It includes such things as packaging, marketing, transportation, storage, and sales. For commercial off-the-shelf (COTS) products and services—that is, those that are ready-made and available to the general public, rather than those that are expressly manufactured for a buyer to custom specifications—this is usually the stage at which buyers can first assess costs and impacts of potential acquisitions.

In addition to manufacturers claims, various aids have been created by different government and private entities to assist both government and private-sector buyers in identifying green COTS products and services. For example, among federal agencies, the Environmental Protection Agency (EPA) provides public access to databases of products assessed as environmental preferable through various criteria.[32] The General Services Administration (GSA) also provides information on green products through its website.[33] The U.S. Department of Agriculture (USDA) has an online catalogue of biobased products listed under its BioPreferred program, which focuses on renewable biobased products.[34] States such as California, Pennsylvania, and Washington have online guides to help buyers identify green products and services, as do some municipal government entities in states such as Minnesota, and organizations such as the National Association of Counties (NACO) and the Northwest Product Stewardship Council.[35] Private- sector organizations such as Green Seal and the Green Electronics Council also provide lists of products.[36] Such aids are discussed further in the "Labeling and Certification" section (see also **Table 3**).

Buyers purchasing custom rather than COTS products and services may have greater capability to ensure the use of green practices in various stages of the life cycle. This may be especially true for large institutions such as the federal government, large states, and major corporations, who are more likely than small purchasers to be able to use their buying power as leverage to achieve such goals.

Operation and Use

A range of impacts may occur at this stage, some of which may be apparent to the purchaser and some not. They include such things as energy use by appliances and vehicles, water use by plumbing fixtures; and release into indoor air of potentially harmful volatile chemicals (called off-gassing) by furniture, furnishings, and building materials.

The degree to which green practices are used at this stage also affects impacts. For example, some users may waste large amounts of paper, ink, and energy through inefficient or unnecessary printing practices; power management of computer systems may not be used effectively; and vehicles may be driven or maintained in ways that compromise efficiency measures.

Products also vary in the length of their useful life. Some, such as disposable cellular phones, may be designed to be discarded after a short period of use. Others, such as modular computers, may be specifically designed to be readily repaired and upgraded rather than discarded when they fail, or in response to technological advances. However, product maintenance and repair may also have environmental impacts, and repair rather than replacement may also affect performance.

Disposal or End of Life

The fate of a product after its useful life has ended may also have significant impacts. Products may be discarded into landfills, refurbished and reused, or recycled—disassembled and parts and materials recovered and reused. Each of those alternatives will have impacts that vary with the kind of product and the processes chosen. While reuse and recycling are generally considered environmentally preferable where possible, the ways in which those processes are carried out can vary significantly in impact. For example, computers and other electronics may be recycled for disposal. It may not be possible to determine before disposal how items will be recycled. Such "e- waste" is often exported.[37] Recycling by recipients may be performed in ways that are dangerous to the health of workers and damaging to the local environment.

Another method used to reduce the impact of disposal is leasing or other forms of "product takeback" at the end of the life cycle. With this approach, a vendor of a product agrees to accept those disposed of by users and recycle them or make them available for reuse. An automobile trade-in is a classic example of this approach, but it is used increasingly with other kinds of products such as electronics.[38]

Green Factors

The main *green factors*,[39] the major determinants of the environmental footprint of a product or service, can be divided into two types, *inputs* and *effects*. Inputs discussed in this chapter include energy, materials, and water; environmental effects include those on ecosystems, health, and waste production. While green factors can be viewed in other ways, the set described here is likely to capture all the major attributes usually considered important.[40]

Inputs

Energy

Energy is often the most prominently considered green factor, because of its economic importance and the environmental impacts of the various methods of generation. Some input of energy is involved at different life cycle stages for all products and services, but its importance may vary substantially across stages for different kinds of acquisitions.

The operation stage is arguably the easiest for which this input can be measured and controlled for any given product, and most attempts to rate or label products with respect to this input focus on such *use energy*. Examples of ratings are Annual Fuel Utilization Efficiency (AFUE)[41] for furnaces, Seasonal Energy Efficiency Ratio (SEER)[42] for air conditioning, and EPA fuel economy estimates for motor vehicles. The most widely known labeling system is probably Energy Star. Programs aimed at energy conservation or efficiency in procurement, including Energy Star, tend to focus on use energy rather than that consumed in other life cycle stages.[43] However, other programs may address energy use at other stages, such as the development of renewable or other green energy sources that may be used by manufacturers.

Energy used during other stages is called *embedded energy* or *embodied energy*.[44] However, embodied energy may be used by some to refer to all energy used during the life cycle, including use energy. This is how it is defined by the Department of Energy (DOE).[45] It may therefore be most helpful to distinguish between the two terms, defining embodied energy as total energy and embedded energy as total energy minus use energy; that is the usage adopted for this chapter. Impacts from life cycle stages other than use and operation may be called embedded impacts.[46]

In some cases, embedded energy may be substantial. For example, concrete exterior walls in buildings may have some advantages over wood, but concrete has much higher embedded energy.[47] However, there appears to be no widely accepted methodology for applying the concept broadly to purchasing decisions, and little information is available for many products. Some observers also argue that considering embedded energy as a separate factor can lead to erroneous conclusions, and that an integrated life cycle assessment is the best way to capture the total energy footprint of a product or service.

One complicating factor for measuring embedded energy is the nature of the energy sources used. For example, if a product with lower embedded energy is manufactured using nonrenewable sources of energy such as fossil fuels, or renewable sources that are produced in ways that have significant environmental impacts, such as on ecosystems, that product may have a greater overall environmental impact than an alternative product with higher levels of embedded energy from more sustainable sources.[48]

Materials

The materials used in a product can contribute substantially to its environmental footprint. The choice and use of materials may have impacts on resource depletion, pollution, embodied energy, and health, among other factors. Green materials may have significant recycled content, be made from renewable biological resources (so-called "biobased" products), or be created with processes that use low amounts of energy and produce low amounts of pollutants. They may also be designed to reduce health risks such as those from off-gassed formaldehyde and other volatile organic compounds (VOCs).

For many kinds of products, for example paper, this factor may exhibit the highest potential impacts at life cycle stages other than use and operation. In such cases, embedded impacts may be of greater importance than use impacts.

Water

Water is a critical resource, and the adequacy of clean freshwater supplies is increasingly a matter of national and global concern.[49] Water is also an important resource at many stages of the life cycles of various products and services. As with energy, its importance varies depending on the product or service and the stage of the life cycle. For example, one U.N. study has estimated that the impact on water resources from production of a pair of leather shoes is 1.6 times that for a ream of paper and 250 times that for a microchip.[50] For none of these is water likely to be a significant resource at the use stage. That is not the case with other products, perhaps the most obvious being plumbing fixtures and irrigation equipment.

Embedded impacts are likely difficult to assess accurately for individual products and services for several reasons, such as regional variations in water supply and the degree to which used water is contaminated and requires processing before reuse. As with energy, most of the focus from the green-procurement perspective has been on the use stage, at which this input may be more readily measured and controlled. For example, the Energy Policy Act of 1992 (EPACT 1992, P.L. 102- 486) established water-use standards for certain plumbing fixtures (42 U.S.C. 6295); EPA's Water Sense labeling program[51] lists water-efficient plumbing fixtures and other products; and rebate incentives for water efficiency have been established in some states.[52] However, attempts have been made to develop measures of what might be called embodied water (by analogy with embodied energy). One such approach is the "water footprint."[53]

Effects

Ecosystem

The term ecosystem is often thought of popularly as referring to natural biological systems largely separate from human civilization. However, in reality humans are embedded in and rely on services provided by ecosystems for such things as freshwater, food and other biobased products, and oxygen, as well as less tangible services relating to quality of life, such as recreation and existence value.[54]

Impacts of products and services on ecosystems therefore have broad application to green procurement. A full assessment would require consideration of a wide range of direct and indirect effects across all life cycle stages and spatial and temporal scales.[55] That is likely to be difficult in many cases, and controversial in at least some.

Health

Several factors can influence the health impacts of products and services. Some impacts are related to the kinds of materials used, such as heavy metals, VOCs, asbestos,[56] or other potentially toxic substances. Other factors, such as radiation, the quality of lighting, effectiveness of climate control equipment in buildings, and ergonomic design of furniture, can also have significant impacts on health.[57]

Such impacts may occur at various stages. For example, the potentially toxic materials used in computers do not generally pose a health hazard for users, but they can pose hazards at both the manufacturing and disposal stages. For some materials, such as metals, impacts from the ways raw materials are originally acquired via mining operations and are processed, such as in smelters, can also pose health risks both to workers and to the public.

Waste

Waste reduction is a common theme of green purchasing programs. Impacts from wastes may occur in several ways. For example, they may be toxic, flammable, or corrosive. According to EPA, more than 46 million tons of hazardous wastes were generated in the United States in 2007.[58]

However, even nonhazardous wastes can be environmentally harmful in ways such as the consumption of energy for disposal, impacts on ecosystems from disposal sites, and resource depletion in cases where recyclable waste is discarded. Impacts may occur at various stages of the life cycle in addition to the disposal phase for a given product or service. In fact, more than two- thirds of U.S. hazardous waste in 2007 was produced by the chemical manufacturing industry.[59]

Evaluation Criteria

The discussion above describes the role of individual green factors, considered over the life cycle, in determining the environmental footprint of a product or service. However, such a description is not sufficient for making informed decisions about whether to acquire a particular product or service. An integrated assessment is needed that uses appropriate criteria. Cost[60] and performance are important in addition to environmental impact (including on human health).[61] Low environmental impact may be of little consequence if a product or service is so costly or performs so poorly in comparison to alternatives that it is not competitive in the marketplace. Other criteria such as availability may also be relevant.

Impact

An overall characterization of impact requires integration of individual impacts from the sets of inputs and effects across all the stages of the life cycle. Reversibility of the impacts must also be taken into account. For example, some chemical pollutants are far more persistent in the environment than others, and some ecosystems recover more quickly from disturbance than others. Some toxins accumulate in the bodies of humans and other species, whereas others are quickly metabolized. Depletion of nonrenewable resources is not reversible.

Cost

While the most obvious component of cost is price, an evaluation of costs across the life cycle includes the total monetary costs to the user of acquiring, operating, maintaining, and disposing of an item. Green purchasing evaluations can expose hidden costs, such as disposal costs, that may otherwise be invisible to those making the purchasing decisions. They may also include indirect costs borne by the purchaser such as medical expenses for hazardous items.

Consequently, a simple assessment of price can be misleading. For example, one study reported that several years ago, the New York Department of Sanitation rejected a proposed purchase of a new fluid for cleaning collection trucks, even though it was environmentally superior, because it cost $0.75 more per gallon than the fluid then in use. However, the price did not take into account other cost factors that would have shown that rather than costing more, the new cleaner would have led to substantial savings.[62]

In addition, externalities may lead to costs borne broadly by the public that would not be reflected in the cost to the purchaser. For example, agroforestry practices that lead to erosion can have adverse impacts on fisheries or water supplies that may not be paid by the producer and therefore would not be passed on to the purchaser in the price of the product.

Performance

Performance refers to how well a product or service accomplishes the functions for which it was acquired. If a reduction in impacts is accompanied by a significant reduction in performance, acquisition may not be cost-effective.

A fairly common perception about green products and services is that they cost more and may not perform as well as standard products. However, there are many cases where green procurement may lead to the acquisition of products and services that are superior in performance to standard ones.[63] This may be especially relevant given that according to a recent survey, most institutional purchasers rate product performance as the most important factor in making purchase decisions.[64]

Life Cycle Analysis

A full examination of options in green procurement involves an integration of the evaluation criteria across the set of green factors and life cycle stages. That involves a life cycle assessment or analysis (LCA), which can be described as a method for analyzing the environmental impacts of something throughout its lifespan, from initial creation through destruction or disposal.[65] This is also known as a "cradle-to-grave" evaluation. A general international standard for such assessments has been developed (ISO 14040 and 14044).[66] Development of at least one U.S. life cycle impact assessment standard (SCS-002) has been initiated.[67]

LCA usually consists of several stages: An initial stage defines the goals and scope of the analysis, including what is to be assessed, the context of the assessment, and the limits of the analysis. That is followed by an inventory of inputs and effects, an assessment of impacts, and an evaluation of the results.[68] Some tools for making such assessments have also been developed, including the Building for Environmental and Economic Sustainability (BEES) tool developed by the National Institute of Standards and Technology (NIST).[69] While it was developed for use in green building projects, the application is broadly applicable; for example, it has been adapted for use in USDA's BioPreferred program. Another example from the building sector is the Athena Institute's EcoCalculator.[70]

An extension of the standard cradle-to-grave life cycle approach is called cradle-to-cradle. It has been described as an attempt to apply the principles of natural biological cycles to product cycles by creating an explicit connection between the end of a product's life cycle and the beginning.[71]

Life cycle analysis can also help ensure that evaluation includes interactions and trade-offs among factors and life cycle stages. A classic example of interaction between inputs and effects is in buildings, where attempts to improve energy efficiency by using products that reduce leakage of air to the outside may permit the build-up of potentially harmful air pollutants, contributing to the phenomenon of "sick building syndrome."[72]

Another example involves incandescent versus fluorescent lightbulbs. While the latter are widely recognized as more energy efficient, concerns have been raised about the extent to which the presence of mercury in the latter counterbalances their advantage in energy efficiency, by posing an environmental hazard from that toxic metal. Some life cycle analyses have found, however, that more mercury is released into the environment over the life cycle of an incandescent bulb, on average, than a fluorescent one, especially in areas where electricity is generated from coal.[73]

An example involving trade-offs relating to the length of the operational stage of the life cycle is the relative impacts of concrete versus wood for the framing and walls of buildings. Concrete can be much more durable than wood and may have other potential benefits, such as fire safety, but the high embedded energy and other impacts of concrete can potentially offset such benefits in low-rise buildings where wood is an option.[74]

Figure 2 depicts a conceptual model of the relationships among the green factors, life cycle stages, and evaluation criteria. In this three-dimensional graphical representation, each element is depicted as a dot on a horizontal matrix, with the criteria forming the third dimension. In making a full assessment using this model, a value for each criterion would need to be determined for each of the 36 elements in the matrix (in the graph, values would be represented by columns of different heights). That would need to be done for each candidate in an acquisition and the results compared to make a decision.

A model of such complexity may be limited in its usefulness in practice, especially for customers making purchase decisions.[75] But it illustrates both some of the central features of green procurement, as discussed above, and some of the difficulties that can be involved in making an accurate assessment. For example, little information may exist in many cases about many of the elements in the matrix.

Also, alternative products and services being considered for a given acquisition may vary substantially in values for different elements and criteria in ways that are difficult to compare. However, such analyses can be used, for example, by trusted third parties who perform such analyses to determine if products meet green certification or labeling criteria, as is discussed in the next section. In addition, assessments may be facilitated by the development and application of standard models and techniques.

Implementing Green Procurement

The discussion above provides a conceptual framework for consideration of policy issues relating to green procurement that might come before Congress, in particular with respect to development, facilitation, and implementation of green procurement in the marketplace, and its adoption and implementation by federal agencies. Several points from the discussion above are particularly relevant to such development and implementation:

- The term *green procurement* covers a range of approaches to acquisition of products and services that have a smaller-than-average environmental footprint.
- The reduction in the environmental footprint of products and services acquired through green-procurement practices may range from marginal to substantial (see **Figure 1**), depending on the specific policies and approaches taken, among other factors.
- A full assessment of a product or service involves an integrated evaluation of several factors across the life cycle of that product or service (see **Figure 2**).
- There are likely to be interactions and trade-offs among different green factors and life cycle stages that would need to be taken into account in any such assessment.
- Information for many of the elements involved may be difficult to assess accurately.
- As a result, assessments may be potentially very complex and may need to be made on substantially incomplete information or through use of proxy measures such as green labels.
- However, because choices are usually comparative—is one choice greener than the others?—full assessments may not usually be required for decisions. Nevertheless, failure to consider a sufficient set of elements may lead to erroneous conclusions.

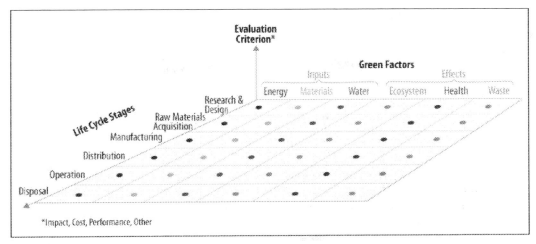

Source: CRS.
Notes: The figure shows a conceptual model of the relationships among the three major dimensions of green procurement as described in this chapter—green factors, life cycle stages, and evaluation criteria. It illustrates that a green procurement assessment involves an integrated evaluation, using appropriate criteria, of green factors (inputs and effects) across all stages of the product or service life cycle. The number of elements and the complexity of the approach implies that in practice, decisions will often need to be made with very limited information about most elements.

Figure 2. Dimensions of Green Procurement and Relationships among Major Elements.

Because of such characteristics and complexities, and other factors such as the ways in which programs, policies, and regulations tend to be developed by both governments and private industry, green procurement initiatives often emphasize particular attributes, such as recycled content, energy efficiency, water use or footprints, waste reduction at end of life, and reduction in hazardous chemicals.[76]

Green procurement may also be complicated by other factors, such as legal requirements or commitments. For example, with respect to international free trade, one author has cautioned that federal "green purchasing regulations must be carefully crafted to avoid the appearance of providing hidden subsidies to U.S. contractors; otherwise, trade disputes will arise through the World Trade Organization."[77]

Several approaches to implementing green purchasing programs have been developed by various government and private-sector organizations, and some examples are discussed below.[78]

Labeling and Certification

One of the most widespread and easiest-to-use approaches to implementation of green procurement is through third-party labeling and certification programs. Examples include

- Energy Star, a joint market-focused energy-efficiency program of EPA and DOE;[79]
- EPEAT, a green electronics label developed by a private, non-profit organization from a recognized technical standard with partial funding from EPA and since adopted by the federal government;[80]

- Green Seal, a private, non-profit environmental certification standard that covers a variety of products and services;[81] and
- Cradle to Cradle (C2C), a private certification program intended to recognize green producers that "close the loop" between disposal and manufacturing, thereby approaching zero-impact.[82]

One organization has identified more than 80 labels relating to green products and services in use in North America and more than 300 worldwide,[83] and the use of labels appears to be on the rise.[84] Such labels and certifications may vary in quality and emphasis. Some, such as EPEAT, may be based on technical standards promulgated or recognized by standard-setting organizations such as the American National Standards Institute (ANSI) or the International Organization for Standardization (ISO).[85] Such standards are developed through a voluntary consensus process involving experts and other stakeholders and are periodically updated. Other labels may be based on criteria developed internally by the labeling entity. If a labeling program has insufficient quality control, it may not reflect the promised environmental performance. In some cases, labels may not reflect any green characteristics, a practice known as "greenwash."[86]

Some labels or certifications focus primarily on a single factor, such as energy use or water, and some involve evaluations of the designated factor only or largely during the operations stage of the life cycle. That is the case, for example, with Energy Star. A focus on a single factor and life cycle stage has benefits in comparative simplicity of evaluation and ease of understanding by potential users, but it could mask trade-offs involving interactions with other factors or effects at other life cycle stages.[87]

Alternatively, labels may be based on a broad set of multifactor or multiattribute standards applied across the life cycle, such as EPEAT and Cradle-to-Cradle. However, such broad, integrative standards may be difficult and time-consuming to develop and keep up-to-date, and as a result, they may apply to relatively few products and may become obsolete. This may be one reason why the federal government uses a variety of certification or labeling systems in its agency-focused initiatives, and why many other organizations use third-party labels and standards as criteria for green procurement decisions, rather than performing the necessary assessments themselves.[88]

Price and Cost Policies

Green products and services may cost more than standard equivalents.[89] In at least some cases, higher acquisition costs may be offset by other savings, as discussed elsewhere in this chapter.[90] Some organizations have implemented policies designed to address such problems. One approach used by various institutions is called price premiums,[91] where institutional policies permit purchasers to pay a somewhat higher price for a product or service that meets green criteria. This approach can help overcome a common problem in the development of green products, namely that economies of scale may keep prices high as long as the market is small. However, the approach can be wasteful because it may reduce price-competitive incentives among providers. Usually, such green price premiums are comparatively small in practice, reportedly ranging from 3%-15% in one study of state and municipal governmental purchasing programs, but they do not appear to be used for federal green procurement programs.[92] The use of this approach also appears to be declining, according to one recent study.[93]

Some organizations address the problem of the kinds of offsets described above by using life cycle cost analyses to determine if savings at other life cycle stages compensate for higher acquisition costs for a product or service. Some state and local governments have also replaced policies based solely on bid prices with what is often called a "best-value" approach.[94] The term may be given different meanings by different institutions, but generally it refers to an effort to take into account, in addition to price, a range of other considerations, such as performance and environmental impact, in making procurement decisions.

Other Approaches

Additional implementation methods used by various organizations may include establishing broad policy goals,[95] use of training and awareness programs, and establishment of cross- functional "green teams" of purchasers, end-users, managers, and environmental experts to facilitate implementation. Another approach is to modify specifications so as to require specific green attributes, such as recycled content or low toxicity, for desired products and services. The way specifications are written may be especially important in avoiding inadvertently precluding green products (for example, specifications that would preclude use of recycled content by specifying "virgin" content).

Measuring Success

A major challenge for both governmental and nongovernmental green procurement programs is how to determine if they are successful in reducing environmental impacts while maintaining or improving cost-effectiveness and performance. Tools have been developed that can be used to estimate the effects of green procurement. Generally, these appear to focus on planning rather than performance assessment.[96]

Success of single-attribute programs, such as increasing recycled content, may be relatively easy to measure. Others, such as those using complex labeling standards, may be more difficult. However, little information may be available even for single-attribute initiatives. For example, in 2001, the Government Accountability Office found that federal agencies could not track the purchases of products with recycled content required by federal law.[97] According to a 2009 survey of governmental and nongovernmental organizations, only 20% of responding organizations with green-purchasing programs quantified the impacts of their programs.[98]

The Green Electronics Council has attempted to improve assessments by requiring EPEAT subscribers to report sales of labeled products. The organization then used those figures, in conjunction with the criteria that labeled products are required to meet, to estimate reductions in toxic materials, solid waste, and energy use over the life cycles of the products. However, those figures are projections, not proven savings.[99]

In 2001, the National Institute of Governmental Purchasing surveyed green procurement among federal and nonfederal government agencies in the United States and Canada.[100] The institute found that one-third of municipal government agencies and one-fifth of nonfederal agencies had green procurement initiatives, but fewer than one in 20 federal agencies had programs. About two-thirds of agencies recognized the Energy Star label, even though energy was listed as a priority by only one-fifth. About 5% of agencies had established goals, but fewer than 20% of those had met them. The obstacles most commonly cited were lack of awareness, conflicting priorities, decentralization of procurement decisions, and inadequate guidance.

A more recent (2009) survey of public- and private-sector organizations suggests a substantial increase in the prevalence of initiatives, but not necessarily improved performance.[101] About three-quarters reported having green procurement policies in 2009, much higher than previous figures, but the organizations also reported that those policies influence fewer than one-fifth of procurement budgets.

The city of Portland, Oregon, has performed several case studies of the costs and benefits of green-purchasing practices by various city bureaus.[102] Among the purchases studied were transportation equipment and supplies, electricity, electronic equipment, inks, and paints. The city found environmental benefits in all cases. In some, such as antifreeze and inks, green products were found to exhibit performance superior to that of standard products. In most cases, the products reportedly reduced the city's acquisition or operational costs as well.

Barriers to Broader Adoption of Green Procurement Practices

As the discussion above implies, several barriers exist to the broader adoption of green procurement practices by both organizations and consumers. Some of those barriers are discussed below.[103]

Incomplete and Imperfect Information

Information barriers tend to fall into two categories: lack of knowledge and incorrect information. The former includes both ignorance about the role of green procurement among potential practitioners and how it can be applied, and gaps in understanding about impacts. The latter includes both misperceptions about what green procurement is and how it can be applied, and misleading claims (greenwash) or other misuse. Training and education of procurement personnel and decisionmakers may be used to help reduce ignorance and misinformation, and regulations and guidelines from government agencies, among other methods, may be used to address misinformation and misuse.[104] Both government and private organizations may also provide information and guidance to consumers, as discussed elsewhere in this chapter. However, considerable research and analysis may be needed to fill current gaps in understanding about impacts.

Lack of Common Standards

The wide range of meanings applied to green procurement and other practices, as well as the absence of broadly accepted assessment and implementation standards, may create both controversy and confusion that can impede the adoption of green procurement and the meeting of policy goals. It may also contribute to the proliferation of green labels of varying accuracy and value.

The apparent success of programs such as Energy Star and EPEAT provide evidence that this barrier can be overcome, and current efforts to develop LCA standards (see the section on

"Life Cycle Analysis" above) may further help to reduce this impediment, but only if such a standard becomes commonly accepted and widely used. Adoption by large industry and government purchasers may facilitate that process.

Real, Perceived, and Hidden Costs

Among the more significant cost barriers are real and perceived cost premiums associated with green procurement.[105] Other barriers include cost externalities—that is, costs that are born by persons other than the purchaser and are not included in the price[106]—and situations in which financial interests of different parties involved are misaligned, which is sometimes referred to as the principal-agent problem.[107] Such barriers may be especially difficult to overcome, except perhaps for misperceptions about cost.

Market and Technical Uncertainties

Uncertainties may include concerns about the stability of market demand for green products and services, which can vary with the costs of inputs such as energy, as well as risks that a nonstandard technology may not be effective in meeting green performance goals, and reluctance to invest in a significant immediate acquisition in a situation such as lighting where technologies are rapidly changing. Those barriers may also be particularly difficult to overcome.

FEDERAL GREEN PURCHASING INITIATIVES

As stated in the introduction to this chapter, federal initiatives relating to green purchasing of products and services can usefully be organized into two categories based on their primary focus:

- agency-focused initiatives intended to facilitate the procurement of green products and services by federal agencies, and
- market-focused programs aimed at the availability of green products and services in the broader economy.[108]

Those categories are not mutually exclusive. Programs that are primarily market-focused, such as Energy Star and BioPreferred, may also affect federal acquisitions.[109] Information on green initiatives developed specifically for federal procurement, such as EPP (see text box below), may be available to the public through an agency website or other source. Agency-focused initiatives may also influence the broader market in at least two other ways. First, nonfederal organizations may adopt some version of a federal initiative, as has happened with EPP.[110] Second, the size of federal procurement operations—about $500 billion per year—means that federal green procurement programs have a significant potential to affect the availability of green products and services in the marketplace, as well as other aspects of the green economy. Given the size and potential impact of the agency-focused initiatives, most of this section addresses those activities rather than market-focused programs.

Market-Focused Programs

Several federal initiatives are either primarily market-focused or have significant market-focused components. Some of those programs are administered by EPA, and some by other agencies. Several are mentioned below in the section on "Preferences for Green Products and Services" and elsewhere in this chapter.[111] While some of the programs are cited in EPA's EPP database,[112] there does not appear to be a comprehensive central federal source for access to such information.

In addition to the guides and labeling programs developed by other federal agencies, the Federal Trade Commission has developed a set of general resources to assist consumers interested in acquiring green products.[113] Perhaps the best known of these resources is the Green Guides.[114] The stated purpose of these voluntary guides is to assist industry in ensuring that environmental marketing claims made for various products and services are not false or misleading.

Agency-Focused Initiatives

Modern federal efforts in green procurement focus largely on acquisition of products. Those efforts arguably began with the enactment of the Resource Conservation and Recovery Act of 1976 (RCRA, P.L. 89-272, as amended), which set requirements for federal purchase of products with recovered or recycled content. Since then, various statutes,[115] regulations,[116] executive orders,[117] policy documents,[118] and agency programs have required or encouraged federal agencies to purchase specific products or types of products because of their environmental attributes,[119] as discussed below.

EPA'S ENVIRONMENTALLY PREFERABLE PURCHASING PROGRAM (EPP)

This program is the only multiproduct, multiattribute, life-cycle-based green procurement program in the federal government, except for certain green building initiatives. Environmentally preferable products and services are characterized as those having "a lesser or reduced effect on human health and the environment when compared with competing products or services that serve the same purpose." This initiative is therefore the closest in concept to green procurement as discussed in this chapter.

EPA's guidance for EPP emphasizes five principles:

1. EPP should integrate environmental, pricing, and performance considerations.
2. Prevention of pollution should be the primary motivation in implementing EPP.
3. EPP should be based on a life cycle approach involving multiple environmental attributes.
4. Agencies should compare environmental attributes of different products and services in determining environmental preferability.
5. Determining environmental preferability requires comprehensive, accurate, and meaningful information about environmental performance.

Established in 1993 in Executive Order 12873, EPP has not been specifically authorized by statute but has been continued by subsequent executive orders and is incorporated in the Federal Acquisition Regulations (48 C.F.R. Subpart 23.7). However, the regulation is arguably one of the weakest of the federal preferences related to green procurement (see text).

Unlike with some single-attribute preferences, EPP does not rely on a specified certification system. Instead, it draws together a variety of federal, private, and international certifications, lists, and standards from which agencies may choose (see **Table 3**).

Sources: Environmental Protection Agency, "Final Guidance on Environmentally Preferable Purchasing for Executive Agencies," Federal Register 64, no. 161 (August 20, 1999): 45810-45858;———, "Environmentally Preferable Purchasing (EPP)," November 7, 2007, http://www.epa.gov/epp;Office of Management and Budget, "Instructions for Implementing Executive Order 13423," March 29, 2007, http://www.fedcenter.gov/_kd/Items/ actions.cfm?action=Show&item_id=6825&destination=ShowItem.

Preferences for Green Products and Services

Some federal government-wide initiatives focus narrowly on specific kinds of products.[120] Others, commonly called preferences, are more general and pertain to types of goods and services that share specified environmental attributes. Currently, there are eight such preferences:

- alternative fuels and alternative fuel vehicles,
- alternatives to ozone-depleting substances,
- alternatives to toxic and priority chemicals,
- biobased products,[121]
- energy-efficient products,[122]
- environmentally preferable purchasing, or EPP (see text box above),
- EPEAT products,[123] and
- recovered or recycled materials.

Currently, agencies *are required*[124] to purchase alternative fuels and alternative-fuel vehicles, and biobased, Energy Star and energy efficient, EPEAT-registered, and recycled-content products, although *acquisitions may be exempt* in specified circumstances. Agencies *must consider* purchasing alternatives to toxic and priority chemicals and ozone-depleting substances, and EPP products and services.

Only EPP and EPEAT attempt to provide an integrative approach to green procurement.[125] The others each address only one or a few attributes. However, EPP and EPEAT, along with alternatives to toxic and priority chemicals, are also the only of those preferences that have no specific basis in enacted statutes.

Most of the preferences listed above apply only to products, despite the fact that about half of federal procurement spending is for services (see **Table 1**). EPP applies to both products and services, although little guidance appears to be available from EPA with respect to the latter. [126] That is especially true for professional, scientific, and technical services, which comprise about half of federal procurement expenditures for services. The energy-efficiency preference also applies to services, specifically for energy-savings performance contracts, in which agencies partner with energy service companies to improve savings in energy use by the agencies.[127] Preferences may also apply to products used by service providers.[128]

Consistent with the current federal emphasis on single attributes, the Obama Administration's Office of Federal Procurement Policy recently described "green purchasing" as including, although not limited to, procurement of "recycled materials; energy-efficient materials or appliances; water saving processes or appliances; environmentally preferable processes, materials, or appliances; and biobased products or materials."[129] EPA, in contrast, characterizes green procurement as essentially synonymous with EPP: "Green Purchasing refers to the practice of preventing waste and pollution by considering environmental impacts, along with price, performance, and other traditional selection factors, when making purchasing decisions."[130]

In addition to implementing the requirements mentioned above, some agencies have broader responsibilities. OMB provides broad guidance through various policy instruments, such as implementing instructions for executive orders,[131] policy letters,[132] and planning documents.[133] Some of those responsibilities are shared with the Office of the Federal Environmental Executive, housed at EPA.

For some kinds of preferences, criteria, which may be standards or even lists of products, are set by specific agencies: DOE and EPA for energy, USDA for biobased, and EPA for recovered materials and certain other types (see **Table 2**). The implementing instructions for Executive Order 13423 state that agencies should use EPA's WaterSense-labeled products and programs to meet water-efficiency requirements, and that DOE will provide guidance on alternative fuels.[134] With respect to hazardous and toxic substances, EPA is designated as the lead agency in the implementing instructions, but there appears to be no specific requirement for development of a list of alternatives. Rather, the emphasis is on reduction in the use of such constituents by individual agencies. EPEAT is a private-sector program but was developed with the assistance of EPA.[135]

There are several government databases containing information about green products, and several of them are publicly accessible. EPA has developed a database listing EPP product and service categories that both agencies and the public can use in making green-purchasing decisions.[136] EPA also has lists of products elsewhere on its website that have been identified under other programs, such as the Significant New Alternatives Program (SNAP) for alternatives to ozone-depleting substances[137] and WaterSense.[138] The USDA has also developed a database listing renewable biobased products currently designated under its BioPreferred label.[139]

GSA provides access to green products and services through its purchasing system available to government agencies.[140] Its online system identifies products registered or identified according to 14 standards and other sets of criteria, including federal requirements such as those for energy, recycled content, emissions, and biobased products, as well as nonfederal programs such as EPEAT and Green Seal.

Green Procurement: Overview and Issues for Congress 53

Table 2. Entities Responsible for Providing Guidance to Federal Agencies for Green Products and Services

Federal Initiative	Agency or Organization			
	EPA	DOE	USDA	GEC
Energy				
Energy Star	X	X		
Federal Energy Management Program		X		
Water				
WaterSense	X			
Materials				
Recovered (recycled) content	X			
Non-ozone-depleting substances	X			
Products with low or no toxic or hazardous constituents	X			
Biobased products			X	
Alternative fuels and vehicles		X		
Multiattribute				
EPEAT				X
Environmentally preferable products	X			

Source: CRS.

Notes: See text for further explanation of individual initiatives. EPA: Environmental Protection Agency; DOE: Department of Energy; USDA: U.S. Department of Agriculture; GEC: Green Electronics Council, a private nonprofit organization.

GSA is also generally responsible for disposing of surplus equipment and other items designated as personal property by federal agencies. However, there appear to be no provisions for GSA to recycle such property if it is not transferred or sold. Property that cannot be used, donated, or sold is returned to the originating agency for disposal.[141] Agencies may have procedures for recycling of such property.

Table 3 lists the kinds of green product and service certifications and labels available through the GSA online catalog or schedule and EPA's EPP database. As the table shows, there are more than 20 designations altogether in the two databases, covering several green factors. Both databases contain integrative or multiattribute labels, but most labels focus on only one or two attributes. Most designations refer to specific standards by various agencies and organizations, but the GSA schedule contains four that are more general: Recycled, Biobased, Non-Toxic, and Environmentally Friendly. Unlike the GSA catalog, the EPA atabase does not list product choices directly but provides electronic links to separate lists, including the GSA schedule.

Table 3. Types of Green Products Listed in GSA and EPA's EPP Databases by Primary Focus

Logo or Icon	Description	GSA	EPA EPP
	Energy		
	Energy Star: Complies with U.S. Department of Energy (DOE) and U.S. Department of Environmental Protection Agency's (EPA) Energy Star Program.	X	X
	FEMP: Meets DOE's Federal Energy Management Program (FEMP) energy-efficiency levels.	X	X
	Materials		
	Recycled: Contains recycled content consistent with EPA guidelines.	X	
	Comprehensive Procurement Guideline compliant: Certified to meet or exceed EPA's Recovered Material Advisory Notice (RMAN) standard.	X	X
	Biobased: Commercial or industrial items utilizing biological products or renewable domestic agricultural (plant, animal, and marine) or forestry materials	X	
	USDA BioPreferred: Commercial or industrial products or ingredients other than food or feed that are composed of biological materials and for which there is not a mature market. The focus is on renewable materials. This logo is not listed separately in the GSA schedule, but some items listed under "Biobased" are designated as BioPreferred.	X	
	PRIME: Included in the U.S. Navy's Plastics Removal in Marine Environment (PRIME) program. Items in the PRIME program require biodegradable packing materials.	X	
	Health and Waste		
	California South Coast Air Quality Management District compliant: Certified to meet or exceed applicable air emissions standards established by the California South Coast Air Quality Management District.	X	
	EPA Design for the Environment: "Best-in-class" wet chemical-based formulations and products through comparisons of toxicity and fate profiles to identify the safest ingredients within a distinct functional class (e.g., surfactants, solvents).		X
	EPA Significant New Alternatives Program (SNAP) Approved: Meets the SNAP criteria requirement as a suitable substitute for ozone-depleting substances.	X	
	NESHAP compliant: "As applied" formulation certified to meet applicable volatile organic compound (VOC) and hazardous air pollutant emission limits as established by the	X	

Green Procurement: Overview and Issues for Congress 55

Logo or Icon	Description	GSA	EPA EPP
	National Emissions Standards for Hazardous Air Pollutant (NESHAP) regulations (40 C.F.R. Part 63). Most items in this category are also listed as "Environmentally Friendly."		
non Toxic	Non-Toxic item	X	
	Multiattribute or Other		
	Environmentally Friendly: Multipurpose icon for items containing specific environmental attributes or features such as water conserving, lead-free, chlorine-free, ozone-safe, low-VOC, etc. as described in the item detail.	X	
ETV	**EPA Environmental Technology Verification (ETV):** Evaluated under the ETV Program and found to be an innovative technical solution, beneficial to human health or the environment.	X	
	EPEAT: Certified as meeting requirements for energy conservation, materials, life cycle management (including disposal), and other factors contained in a standard developed and promulgated by the Institute of Electrical and Electronic Engineers (IEEE).	X	
	GreenSeal Seal of Approval: Certified to meet or exceed voluntary standards for environmental preferability as established by GreenSeal.	X	X
	McDonough Braungart Design Chemistry (MBDC): Certified as designed for cradle-to-cradle life cycle sustainability based on environmental chemistry, material flows management, and industrial and architectural design.		X
SCS	**Scientific Certification Systems (SCS):** Certified for environmental, sustainability, food quality and food purity claims.		X
Various	**International:** Includes labels from programs in Canada, European Union, Japan, and Nordic countries.		X

Sources: General Services Administration, 2009 GSA Global Supply Catalog: Environmental Products, 2009, http://apps.fss.gsa.gov/cmls/search.cfm?keyword=2009%20GSA%20Global%20Supply%20Catalog:%20Environmental%20Products; Environmental Protection Agency, "Environmentally Preferable Purchasing (EPP)"; Department of Agriculture, "BioPreferred"; Green Electronics Council, "Welcome to EPEAT"; MBDC, Cradle to Cradle Certification Program, Version 2.1.1; Scientific Certification Systems, "Scientific Certification Systems," 2009, http://www.scscertified.com/index.php; U.S. Navy, "Plastics Removal in the Marine Environment," 2006, http://www.navy.mil/oceans

Note: Primary focus (Energy, Materials, and so forth) is a CRS estimate based on the description. Some designations may also have focuses related to more than one green factor; for example, PRIME materials are also ecosystem-related and types involving air emissions are also waste-related. Designations included in the table are only those explicitly listed in the relevant GSA catalogs and EPA's Environmentally Preferable Purchasing database; for example, EPA's WaterSense label is not currently included in either database. Logos and icons are from the sources.

The Office of the Federal Environmental Executive also produces a database of green products as a tool for federal agencies to use in procurement. It contains green products designated by EPA, USDA, and DOE, with respect to Energy Star, FEMP, WaterSense, recycled and biobased content (including BioPreferred), alternative fuels, and SNAP, as well as EPEAT, with links to relevant sources.[142]

Federal Performance in Green Procurement

Gauging the degree to which the various federal green-purchasing requirements have been applied is difficult. Agencies vary greatly in the amounts of procurement they engage in (see, for example, **Table 4**), with the Department of Defense (DOD) accounting for almost three-quarters of total federal procurement spending in FY2008, more than 15 times as much as the next highest-spending agency, DOE.

Table 4. Federal Procurement Spending for FY2008 by Department or Agency for Selected Industries

Department/Agency	Amount ($ Billions)			
	Total	Transportation Equipment	Information Technology	Building Construction
Defense	396.2	109.2	19.1	36.1
Energy	24.8	0.0	0.4	0.2
National Aeronautics & Space Administration	15.0	1.8	0.9	0.2
Veterans Affairs	14.8	0.0	1.1	1.7
Homeland Security	14.2	0.7	3.1	0.3
Health and Human Services	13.7	0.0	2.2	0.1
General Services Administration	12.7	1.3	2.4	1.3
State	6.1	0.4	0.7	0.8
Justice	6.1	0.0	1.4	0.2
Agriculture	5.3	0.0	0.7	0.0
Transportation	4.8	0.0	0.7	0.1
Treasury	4.5	0.0	1.2	0.0
Interior	3.8	0.0	0.7	0.2
Agency for International Development	3.5	0.0	0.0	0.0
Commerce	2.5	0.0	1.1	0.1
Labor	1.8	0.0	0.3	0.0
Education	1.4	0.0	0.3	0.0
Environmental Protection Agency	1.4	0.0	0.2	0.0
Other	5.2	0.1	2.5	0.3
Grand Total	**537.8**	**113.6**	**39**	**41.6**

Source: Analysis by CRS of data from Federal Procurement Data System (FPDS-NG), October 2009, https://www.fpds.gov.

Note: Industries correspond to those in **Table 1**, except for information technology, which includes both products and services relating to computers, peripheral devices, and related items. Slight differences in totals in the two tables are the result of minor differences in the way the reports on which they are based were generated. **Table 4** includes only FPDS-NG data from FY2008, because they are expected to be less subject to change at the level of disaggregation shown than the more recent FY2009 data; however, total procurement expenditures in the database for the agencies listed were similar for the two years.

FEDERAL GREEN IT PROCUREMENT

The information technology (IT) used by the federal government has a significant environmental footprint:

- Federal agencies invest about $70 billion in computers and related resources each year. About $10 billion of that amount is spent on products and about $30 billion on services, with the rest invested in agency personnel and other costs.
- IT is an increasingly important component of other products and services, such as telephony, television and radio, transportation, and buildings.
- Almost 10% of electricity use in federal office buildings is for electronics.
- It takes eight gallons of water to make one computer chip.
- IT often contains lead, mercury, and other toxic chemicals.
- Most federal computers are replaced after only three years, less than half their design life.
- Agencies dispose of about 750,000 computers and monitors each year.
- Many recycling operations for computers and other "e-waste" items are overseas and are often not subject to effective environmental regulation.

Several federal policies and programs have been established to promote the greening of federal IT. Among them are the following:

EPEAT. For the types of IT products registered with the Electronic Product Environmental Assessment Tool (EPEAT) label, at least 95% of agency acquisitions are to consist of registered products. Products are evaluated according to criteria in a technical standard from the Institute of Electrical and Electronic Engineers (IEEE). To qualify, products must meet requirements for energy conservation, materials, life cycle management (including disposal), and other factors. At present, only desktop and laptop computers and monitors are registered. Standards for servers, mobile devices, and other products are in development. Also, the EPEAT requirements relate only to product acquisition, not services, which comprise more than half of federal IT procurement.

Energy Efficiency. Agencies are to procure energy-efficient products, either Energy Star or FEM P-designated, enable the energy-saving features on all Energy Star computers and monitors, and acquire products with low standby power requirements.

Life Cycle. Agencies are to extend the useful life of electronics for at least four years whenever possible and are to ensure that they are then reused, donated, sold, or recycled in an environmentally sound manner.

Electronic Stewardship Plan. Each agency must have a plan describing how it will meet the goals above and report on progress in meeting those goals.

Federal Electronics Challenge. This EPA program assists agencies in meeting green IT goals and provides recognition for success. Agencies must participate in the challenge or implement an equivalent program. Sixteen agencies participated in 2009, accounting for about 90% of federal IT procurement spending. For FY2008, participants reported an 88% success rate in acquiring EPEAT-registered products; 98% reuse, recycling or selling of IT taken out of service; and an average useful life of 45 months; but only 38% of computers with Energy Star power-management features enabled.

Cloud Computing. Recent federal initiatives in cloud computing—the use of shared, networked computer resources for data storage, applications, and other IT services—also hold significant promise for greening federal IT through reduction in energy use and hardware requirements.

Sources: Cate Berard, "Federal Electronics Stewardship 101" (presented at the 2009 Federal Environmental Symposium, Bethesda, MD, June 16, 2009), http://www.fedcentr.gov/_kd/go.cfm? destination=ShowItem&Item_ID=12426; A.K. Chapagain and A.Y. Hoekstra, Water Footprints of Nations, Value of Water Research Report Series No. 16 (UNESCO-IHE, November 2004), http://www.waterfootprint.org/Reports/Report16Vol1.pdf; Green Electronics Council, "Standards Development Process and Participation," 2009, http://www.epeat.net/StandardsDevelopment.aspx; General Services Administration, "Apps.Gov," https://www.apps.gov/cloud/advantage/main/start_page.do; Peter Mell and Tim Grance, "The NIST Definition of Cloud Computing v 15," October 7, 2009, http://csrc.nist.gov/ groups/SNS/cloud-computing/cloud-def-v 1 5.doc; Office of Management and Budget, "Federal IT Dashboard," June 12, 2009, http://it.usaspending.gov/;————, "Instructions for Implementing Executive Order 13423," March 29, 2007, http://www.fedcenter.gov/_kd/Items/actions.cfm?action=Show&item_id=68 25&destination= ShowItem; Analysis by CRS of data from Federal Procurement Data System (FPDS-NG), October 2009, https://www.fpds.gov.

DOD's dominance was also reflected in specific areas of substantial procurement spending that have been the subject of federal green-procurement initiatives—transportation equipment (e.g., alternative fuels[143]), information technology (electronic stewardship[144]), and building construction (green building[145]). Agency totals for procurement spending for those three industry groups are also presented in the table. Among them, the most comprehensive federal initiatives currently for green procurement are for information technology (see text box above).

OMB requires agencies to report annually on their activities in green procurement. However, those reporting requirements appear to be largely qualitative.[146] There does not appear to be any broad requirement for or practice of federal agencies reporting the extent of their green purchases.

However, quantitative reporting is required for the recycled content and alternative fuels preferences. For the former, reports can be generated through the Federal Procurement Data System (FPDS-NG).[147] Such purchases were reported as totaling $35.9 billion in FY2008, comprising about 7% of total procurement expenditures. Less than 1% of those purchases, totaling $333 million, were reported as exempt from the requirement because of performance, price, or time constraints.[148] Even those data are limited, since many relevant purchases are not included because they fall below the reporting threshold or for other reasons.

For alternative fuels, DOE requires agencies to report annually on vehicle acquisitions and fuel consumption.[149] According to DOE, agencies largely met or exceeded the requirements for FY2007, acquiring almost twice the required number of alternative fuel vehicles, and exceeding the target for alternative fuel consumption by almost 20%.[150]

> ## ENVIRONMENTAL MANAGEMENT SYSTEMS (EMS)
>
> Executive Order 13423 requires that agencies implement environmental management systems "at all appropriate organizational levels" and use them as "the primary management approach for addressing the environmental aspects of internal agency operations and activities."
>
> An EMS is a specified set of processes and practices for assisting an organization in reducing the environmental impacts of its activities. The processes are designed to incorporate continuous improvement and involve planning, implementing, assessing, and revising. A set of international standards have been developed to assist organizations in establishing such systems.
>
> OMB's implementing instructions for the executive order specify that "sustainable acquisition" is to be among the goals addressed by the EMS. Agencies are to report annually on implementation.
>
> Sources: Executive Order 13423, "Strengthening Federal Environmental, Energy, and Transportation Management," Federal Register 72, no. 17 (January 26, 2007): 3919-3923; International Organization for Standardization, "ISO 14000 essentials," 2009, http://www.iso.org/iso/iso_14000_essentials; Office of Management and Budget, "Instructions for Implementing Executive Order 13423," March 29, 2007, http://www.fedcenter.gov/_kd/Items/actions.cfm?action=Show&item_id=6825&destination= Federal Facilities Environmental Stewardship and Compliance Assistance Center, "Environmental Management Systems (EMS)," November 17, 2009, http://www.fedcenter.gov/programs/EMS.

A January 2009 report to Congress from the Office of Federal Procurement Policy describes agency compliance with the green purchasing requirements in E.O. 13423.[151] As part of that compliance, OMB requires agencies to have "formal, written, documented green purchasing plans, policies and/or procedures for the implementation of the statutory and executive order requirements to purchase green products and services."[152] According to the 2009 report, all agencies have such plans in place and have demonstrated compliance in "representative acquisitions."

Plans for at least some agencies are comprehensive. For example, DOD has developed a department-wide strategy that requires each organization involved in procurement to have a green procurement plan and to integrate or coordinate that plan with the environmental management system (EMS) that agencies were also required to establish under that and other executive orders (see text box above).[153] The Defense Logistics Agency also provides both a database of products analogous to that of GSA and a reporting capability for green procurement that appears to be more extensive than that available through FPDS-NG.[154]

DOE guidance on green procurement also stipulates that plans should be integrated with the agency's EMS.[155] The guidance also lays out reporting requirements.

The January 2009 report[156] cited above stated that most agencies were monitoring compliance and developing corrective actions where necessary. However, the report also recognized the difficulties with collection of data on green purchasing but stated that new requirements are being drafted for the FPDS-NG that would permit routine capture of green-purchasing data.

Implementation and performance of green procurement programs are also tracked by OMB through agency submissions for performance scorecards on energy, transportation, and environment. The environmental submissions include information on EMS, green purchasing, green buildings, electronic stewardship, compliance plans, and implementation strategies. Agencies can receive one of three different scores—green (denoting success), yellow (mixed performance), or red (serious flaws). [157] Of the 25 agencies in the January 2009 summary, only four scored green for their overall environmental performance—Commerce, DOE, EPA, and Treasury. Nine scored yellow, and 11 scored red, including DOD and GSA. [158] GSA's scorecard noted deficiencies in green procurement auditing, implementation of an agency-wide plan, and compliance monitoring. Specific information relating to the scorecard rating was not available for DOD.

ISSUES FOR CONGRESS

The discussion above raises several policy issues that Congress may consider in examining federal efforts in green procurement and determining whether any legislative initiatives might be appropriate. Questions raised include the following:

- What, if any, are the most useful and appropriate policy goals for green procurement?
- Are the means by which different green-procurement preferences, programs, and other initiatives have been established the most appropriate for meeting policy goals?
- How effectively are agency implementation and performance of green procurement being assessed?
- How successful are current programs and initiatives at meeting policy goals?
- Are policies on the acquisition of green services sufficient?
- Are the preferences and the methods of implementing them sufficiently harmonized and integrated?
- Are there significant gaps in the various federal preferences for types of green products and services?
- Are there implementation methods not currently used by the federal government that should be considered?
- Is training of procurement officials sufficient?

The issues are explored and further characterized below.

Green Procurement Policy Goals

Green procurement requirements and other initiatives can be used to help meet several policy goals. Perhaps the most obvious and direct is the reduction of the environmental footprint of federal activities with respect to a set of green factors such as those described in this chapter. Related to that is cost savings through reduction of inputs such as energy, water, and materials. Another goal could be to use procurement leverage or market-focused initiatives to actively facilitate the development of green industries and other aspects of the green economy. Also, federal programs could help provide better ways to meet such goals, for example through facilitating the maturation of life cycle assessment methods and the development of standards.

Some broad goals are contained in executive orders and related policy documents.[159] Congress might find useful an integrated assessment of such goals and options for achieving them. Also, some observers may question more broadly whether green procurement is an appropriate focus for federal policy. A discussion of the complex and often longstanding controversies about the desirability of green policy goals in general is beyond the scope of this chapter, but an assessment of green procurement policy might include an evaluation of the appropriateness of the goals Congress has already established in various statutes, as well as those specifically addressed in executive orders issued by Presidents Clinton, George W. Bush, and Obama, in addition to new policy proposals.

Also, given the lack of a commonly accepted characterization of what green procurement involves, especially in the context of other concepts and approaches such as sustainable procurement, an examination of green procurement and other relevant concepts, including whether they should be defined in legislation, might also be considered as part of an examination of policy goals.[160] In addition, an assessment might include examination of how green procurement policy goals interact with other goals of federal procurement policy.

Green Procurement Policy Instruments

The particular method used by the federal government to establish a green-procurement policy or requirement can have implications both for the degree to which it will be implemented and the ease with which it may be modified or revoked. In general, statutory requirements such as those for recycled content and alternative fuels are likely to be more fully implemented than a conditional preference such as EPP established by executive order but not by an enacted statute.

Among the various methods available, policy documents[161] provide the most flexibility but the least potential stability, as the executive branch can modify or revoke them at any time without necessarily considering congressional input or public comments. Statutes, in contrast, provide significant stability because Congress must enact legislation to revoke or modify them.

However, the comparative difficulty of changing statutes may also make it more difficult for government to adjust policies established by them when external conditions change. That may be especially the case if the statute establishes a fixed design standard rather than a flexible performance requirement. For example, a statutory requirement to use a specific kind of alternative fuel might impede adoption of emerging alternatives that may be substantially better for meeting energy-policy goals. However, a specific design standard is often simpler to implement than a performance standard.

Executive orders and regulations based on them may be instituted, revised, or revoked without congressional action, but Congress has the opportunity to review executive orders and regulations before they take effect, and the public generally has the opportunity during the rulemaking process to comment on proposed regulations or proposed modifications to existing regulations. Those two instruments therefore arguably fall somewhere between policy documents and enacted statutes in terms of their flexibility and stability, but unless they are specifically based on enacted statutes, they also give primary authority to the executive branch for any resulting green procurement requirements, policies, and programs.

Given the broad range of green-purchasing preferences (see the section on "Preferences for Green Products and Services") and the variation in the strength of their applicability, stability, and flexibility, Congress may opt to examine whether any modifications would be desirable. In some cases, Congress has chosen to create statutory authority for an existing preference, as when it provided a statutory basis in the Farm Security and Rural Investment Act of 2002 (P.L. 107-171) for preferences for biobased products that already existed under executive orders and regulations.

Performance Assessment

The methods currently used to track agencies' performance in green procurement, including those contained in environmental scorecards and reports to Congress, focus largely on processes, not outcomes. They tend to examine whether agencies have programs and policies in place rather than quantitative measures of the results of those programs and policies. Also, to the extent that reporting requirements are fragmented, it may be difficult for Congress to assess the degree to which agencies are successfully engaging in integrated green-procurement activities that adequately address interactions among the various green factors across the life cycles of products and services.

Agency environmental management systems (see EMS text box above) could provide a mechanism for integrated performance tracking and assessment, but only if agencies collect appropriate life cycle data on relevant acquisitions, including procurement amounts, useful life, and end-of-life management. OMB has directed agencies to consider life cycle costs in procurement planning,[162] but the degree to which the information they collect for that purpose would need to be supplemented to be useful for EMS analyses is not clear.

Program Evaluation

While performance assessment is important for determining how well existing programs are being implemented, it does not provide a method for examining how well a program or a set of programs is meeting intended policy goals. Some entities have performed studies about outcomes from programs relating to federal green purchasing preferences,[163] but there does not appear to have been any broader, independent examination[164] of the success of those programs and initiatives in meeting broader policy goals such as those discussed above. Such an examination might be useful to Congress in determining whether programs should be modified, expanded, or replaced.

Acquisition of Services

Roughly half of all federal procurement spending is for services.[165] However, many of the green purchasing preferences apply only to products. For example, while recovered-content products are preferred in contracts for goods valued at over $10,000, they are preferred in contracts for services only for solid waste management.[166]

Such restricted applicability may limit the ability of the federal government to use its purchasing power to leverage the availability and cost-effectiveness of green products and services in the marketplace, while broader use of such leveraging may help stimulate the development of green industries, to the extent that is a policy goal. However, the cost and effectiveness of such efforts is not clear.

Some recent proposals are seen as having had the potential to extend the product preferences to additional service contracts, although none have yet been so extended. The George W. Bush Administration proposed substituting green products for equivalent non-green products in the "performance" of service contracts.[167] That proposal could have required contractors to use green products when doing work for the government even when that work did not involve the supply or delivery of products to federal agencies or facilities. However, the proposal was never finalized.

President Obama's Executive Order on "Federal Leadership in Environmental, Energy, and Economic Performance" (E.O. 13514) contains provisions that could potentially have a similar effect. However, it is unclear whether they will be interpreted to such effect in either regulations or other policy implementation, and Congress does not appear to have provided any general guidance to agencies on the applicability of green purchasing preferences and programs to various services.

Integration and Harmonization of Green Procurement Initiatives

The federal approach to green procurement is arguably largely piecemeal and fragmented. The emphasis has largely been on single-attribute preferences, with the exception of EPP, which was designed in a way that could permit it to serve as a framework for an integrated, life cycle approach to federal green procurement.[168] However, it has been implemented as just one among the various preferences, and a comparatively weak one. It has no specific basis in enacted statute and no requirement for formal determination of exemptions.

Additionally, there appears to be significant ambiguity about which type of green product or service agencies should procure in situations where multiple types could meet their needs. For example, the FAR requires agencies to acquire recovered-content products instead of biobased ones when both types would meet agency needs.[169] However, no similar guidance exists for the other types of preferred products and services discussed in this chapter. That leaves agencies without guidance in determining whether, for example, they should procure Energy Star or FEMP-designated products, or recovered-content or environmentally preferable products. Such issues could potentially be addressed through an interagency mechanism such as the Interagency Sustainability Steering Committee established under E.O. 13423 and its working group on acquisition and materials management. An examination of the role of these bodies may be useful in determining how best to effect improved integration and harmonization of the various initiatives.

Possible Gaps in Current Preferences

While EPP is arguably broad enough in concept to include all the factors involved in green procurement, its current limitations in comparison with single-attribute preferences have led to some gaps in practice in the coverage of the green factors and life cycle approach discussed in this chapter. Three gaps that are particularly notable are

- the focus of the energy-efficiency preference on use (or operational) energy rather than life cycle energy,
- the lack of an explicit preference with respect to water use and footprints,[170] and
- the absence of specific green requirements for end-of-life management for government property such as information technology.

Also, EPP is among the weakest of the preferences, since it does not require any specific percentages or amounts of environmentally preferable acquisitions and is not accompanied by any requirement for contractor certification.

At present, there is also no single, generally accepted, multiattribute life-cycle-based certification or labeling system that procurement officials can use for all green products and services, as, for example, EPEAT provides for information technology. There is not even a single, well-coordinated list of green products for federal agencies to use, but rather several lists maintained by different agencies. In the absence of such resources and stronger requirements, significant gaps in preferences are liable to persist and the federal green procurement process is likely to remain relatively complex and opaque. However, given the complexities of the procurement process, the potential for success and possible disadvantages of bridging those gaps are not clear and might be difficult to ascertain.

Additional or Different Implementation Mechanisms

Some foreign, state, and local governments have mechanisms for implementing green procurement that the federal government does not use. To the degree that Congress finds the current level of green procurement by federal agencies inadequate, or agencies have difficulties implementing green procurement, Congress could promote additional or different implementing mechanisms. Doing so could including granting agencies additional legal authorities.

Training of the Acquisition Workforce

Given concerns that have been expressed about the capabilities of the acquisition workforce,[171] Congress could explore whether the current workforce can adequately implement existing or proposed requirements and policies pertaining to green procurement.[172] Certain requirements, most notably those involving environmentally preferable products, may be difficult for the existing workforce to implement because agencies must consider multiple attributes of products when determining which product to purchase. This is a more complex determination than ascertaining whether a product possesses a single attribute, such as energy efficiency, and agencies might need information about vendors' production methods to make such determinations. Proposed requirements focused on products' sustainability or life cycle costs could be difficult to implement for similar reasons.

End Notes

[1] See, for example, Joel Makower, *State of Green Business 2010* (Greener World Media, February 2010), http://stateofgreenbusiness.com/node/1.

[2] Reported spending is total dollars obligated and in FY2008 represented an increase of 15% over FY2007 and 25% over FY2006 (data generated by CRS from the Federal Procurement Data System, FPDS-NG, https://www.fpds.gov). However, spending in the table for FY2009 was slightly lower than FY2008 (0.3%). FPDS-NG lists all actions valued at over $3,000 on executive-branch procurement contracts. It is the only comprehensive source for data of this kind, but commentators have repeatedly questioned the accuracy and completeness of its data. For more about FPDS-NG and concerns about the quality of its data, see CRS Report RL3471 8, *The Federal Funding Accountability and Transparency Act: Implementation and Proposed Amendments*, by Garrett Hatch.

[3] Construction is often treated separately from services. However, for simplicity of exposition, the term *products and services* as used in this chapter means products, services, and construction unless otherwise indicated.

[4] See 48 C.F.R. § 23.203 and 42 U.S.C. 17091.

[5] However, in most cases, it is not specifically called "green procurement" in those statues and executive orders, as discussed later in this memorandum.

[6] See, for example, CRS "Issues in Focus" on Greenhouse Gas Policy, and Chemicals in Commerce, and CRS Report R40168, *Alternative Fuels and Advanced Technology Vehicles: Issues in Congress*, by Brent D. Yacobucci.

[7] See, for example, CRS "Issues in Focus" on Government Procurement.

[8] That characterization is similar in concept to one often used to describe green building. See CRS Report R40 147, *Issues in Green Building and the Federal Response: An Introduction*, by Eric A. Fischer.

[9] For example, a programmable thermostat for a furnace or air conditioner can provide significant energy savings, but purchasing the device is not sufficient. The programming features must be used appropriately, or no energy savings will result.

[10] However, in some cases, businesses may attempt to promote nongreen practices as green. This practice, known as "greenwash," is more likely to be an issue for products or services that are not certified under an established green program or label such as Energy Star. However, it can sometimes be an issue even for such an established program. See, for example, Office of Inspector General, *The Department's Management of the ENERGY STAR Program*, DOE/IG-0827 (Department of Energy, October 2009), http://www.ig.energy.gov/documents/IG-0827-508.pdf.

[11] Some observers make this point by claiming that green procurement is more appropriately called greener procurement.

[12] Sustainability may also have various meanings in different contexts. A widely cited definition, in the context of development, is "meet[ing] the needs of the present without compromising the ability of future generations to meet their own needs" (United Nations, *Report of the World Commission on Environment and Development: Our Common Future*, A/42/427, June 1987, http://www.un-documents.net/wced-ocf.htm; commonly known as the Brundtland Commission Report). Executive Order 13514, "Federal Leadership in Environmental, Energy, and Economic Performance," defines it as meaning "to create and maintain conditions, under which humans and nature can exist in productive harmony, that permit fulfilling the social, economic, and other requirements of present and future generations" (74 Fed. Reg. 52126 (October 8, 2009)).

[13] Sec. 433 modifies 42 U.S.C. § 6834(a)(3) to require that buildings "be designed so that the fossil fuel-generated energy consumption of the buildings is reduced, as compared with such energy consumption by a similar building in fiscal year 2003" by a percentage that increases every five years, culminating in a 100% reduction (i.e., zero net energy use) by 2030.

[14] Of course, if zero-impact became the industry standard practice, it would no longer be considered green under a strictly relative definition. Rather, that would be limited to practices that enhanced green factors, such as by removing excess greenhouse gases from the atmosphere or restoring lost ecosystem functions from a landscape.

[15] See, for example, ASQ, "Continuous Improvement," 2009, http://www.asq.org/learn-about-quality/continuous-improvement/overview/overview.html.

[16] See, for example, International Organization for Standardization, "ISO - Standards development processes - How are standards developed?," 2009, http://www.iso.org/iso/standards_development/processes_and_procedures/how_are_standards_developed.htm. Note that *standards* in this sense differs in meaning from *standard practices* or *industry standards*. The latter refer to practices and characteristics that are usual or typical for a particular sector or industry, whereas the former usually refers to a *documentary standard*, which is a consensus document, approved by a recognized organization, that provides "rules, guidelines, and characteristics for activities or their results" (Christine R. DeVaux, *A Guide to Documentary Standards*, NISTIR 6802 (National Institute of Standards and Technology, December 2001), http://ts.nist.gov/Standards/Conformity/upload/ir6802.pdf).

[17] Section 12(d) of the National Technology Transfer Advancement Act of 1995 (P.L. 104-113), as amended, (15 U.S.C. § 272 note) states, "all Federal agencies and departments shall use technical standards that are developed or adopted by voluntary consensus standards bodies, using such technical standards as a means to carry out policy objectives or activities determined by the agencies and departments," except where they would be "inconsistent with applicable law or otherwise impractical." Office of Management and Budget Circular A-1 19 provides guidance for implementing this provision.

[18] For example, a provision in EISA relating to energy-efficiency standards for industrial equipment requires the Secretary of Energy to "establish an amended uniform national standard for the product at the minimum level specified in the amended ASHRAE/IES Standard 90.1" (§ 305(b)(2)).

[19] For example, California requires that "All carpet purchased by ... state agencies shall meet the NSF/American National Standards Institute 140-2007 standard at its PLATINUM level" (State of California, "Carpet Purchases to Meet Environmentally Preferable Criteria," Management Memo MM 1—01, State Administrative Manual, December 31, 2009, http://www.documents.dgs.ca.gov/osp/sam/memos/MM10_01.pdf).

[20] DeVaux, *A Guide to Documentary Standards*. Performance standards focus on functional requirements, whereas design standards focus on specific characteristics. Although both are useful and often form elements of a particular documentary standard, the former are generally preferred where applicable because they are less restrictive and are thought more likely to encourage innovation. However, assessment of conformance may be more complex for performance standards.

[21] For example, that is how the Responsible Purchasing Network defines the term (Responsible Purchasing Network, "Factsheet: Buying Green on Cooperative Contracts," 2009, http://www.responsiblepurchasing.org/includes/ cooperative_contracts.pdf). The federal definition of environmentally preferable products and services is those having "a lesser or reduced effect on human health and the environment when compared with competing products or services that serve the same purpose" (see the text box on EPA's Environmentally Preferable Purchasing Program (EPP) below).

[22] For example, the state of California has implemented an EPP initiative (California Integrated Waste Management Board, "Environmentally Preferable Purchasing (EPP) Home Page," January 1, 2010, http://www.ciwmb.ca.gov/EPP), and at least 90 governmental and nongovernmental organizations, including corporations, in the United States have reported implementing EPP policies (Responsible Purchasing Network, *Responsible Purchasing Trends*, 2009, http://www.responsiblepurchasing.org/publications/trends_2009.pdf).

[23] In fact, the federal EPP initiative is often treated as one of several kinds of green procurement in federal documents, as discussed later in this chapter. Nevertheless, green procurement as described here does have more characteristics in common with EPP than with other federal initiatives—specifically, both incorporate relative assessments and take an integrated, multifactor, life cycle approach.

[24] For example, the City of Boulder, Colorado, has defined an environmentally preferable product with respect to specific attributes, as "a material or product which is durable, repairable, reusable, or recyclable; has a minimum of packaging, toxic content or chemical hazard potential; is resource or energy efficient in any or all phases of its manufacture, use, and disposal; or in its use or disposal minimizes or eliminates the City's potential environmental liability" (Scot Case, *Environmental Purchasing Policies 101: An Overview of Current EPP Policies* (Commission for Environmental Cooperation and the North American Green Purchasing Initiative, March 25, 2004), http://www.cec.org/files/PDF//NAGPI%20Policy%20Paper2e.pdf).

[25] Environmental Protection Agency, *Energy Star—the Power to Protect the Environment through Energy Efficiency*, July 2003, http://www.energystar.gov/ia/partners/downloads/energy. Green Electronics Council, "Welcome to EPEAT," 2006, http://www.epeat.net/default.aspx. EPEAT is an acronym for the Electronic Product Environmental Assessment Tool.

[26] See, for example, Environmental Protection Agency, "Greening Your Purchase of Cleaning Products: A Guide For Federal Purchasers," August 10, 2009, http://www.epa.gov/epp/pubs/cleaning

[27] A classic example is in the construction of buildings, where it is widely recognized that making a building green by retrofitting is more expensive and less effective than incorporating green features in the design phase (see CRS Report R40 147, *Issues in Green Building and the Federal Response: An Introduction*).

[28] One example is the Toxics Use Reduction Institute (TURI) at the University of Massachusetts (http://www.turi.org), which does research to find alternatives to toxic chemicals. See also Daniel Sarewitz, "A tale of two sciences," *Nature* 462 (November 19, 2009): 566. See also Environmental Protection Agency, "Design for the Environment (DfE)," November 10, 2009, http://www.epa.gov/oppt/dfe.

[29] See, for example, National Oceanic and Atmospheric Administration, "Fisheries Interactions/Protected Species Bycatch," 2009, http://www.nmfs.noaa.gov/pr/interactions. *By-catch* refers to the harvesting by fishing operations of unwanted species, such as dolphins or porpoises being caught in tuna nets, or sea turtles by shrimp trawlers.

[30] See, for example, Environmental Protection Agency, "Mining Waste," January 13, 2009, http://www.epa.gov/osw/ nonhaz/industrial/special/mining.

[31] Environmental Protection Agency, *Lean Manufacturing and the Environment: Research on Advanced Manufacturing Systems and the Environment and Recommendations for Leveraging Better Environmental Performance*, EPA100-R03-005, October 2003, http://www.epa.gov/lean/leanreport.pdf.

[32] Environmental Protection Agency, "Environmentally Preferable Purchasing (EPP)," November 7, 2007, http://www.epa.gov/epp.

[33] General Services Administration, "Environmental Products Overview," June 29, 2009, http://www.gsa.gov/Portal/ gsa/ep/channelView. do?pageTypeId=171 10&channelPage=%2Fep%2Fchannel%2FgsaOverview.jsp&channelId=- 24454.

[34] U.S. Department of Agriculture, "BioPreferred," 2009, http://www.biopreferred.gov/Default.aspx.

[35] California Department of Resources Recycling and Recovery, "Green Guide: California Greenin'," 2010, http://www.calrecycle.ca.gov/EPP/Procurement/GreenGuide/default.htm; Commonwealth of Pennsylvania, "Green Procurement," n.d., http://www.portal.state green_procurement_hom e/49 1348; Department of Ecology, State of Washington, "What Is Environmentally Preferable Purchasing?," n.d., http://www.ecy.wa.gov/beyondwaste/epp.html; Solid Waste Management Coordinating Board, "Environmentally Preferable Purchasing Guide," April 25, 2008, http://www.rethinkrecycling.co m/government/eppg; National Association of Counties, "NACo's Environmental Purchasing Starter Kit," 2009, http://www.countyengineers.org/Template.cfm?Section=New_Technical_Assist ance&template=/ContentManagement/ ContentDisplay.cfm&ContentID=16814; Northwest Product Stewardship Council, "Northwest Product Stewardship Council," September 22, 2009, http://www.productstewardship.net/index.html.

[36] Green Seal, "Green Seal," n.d., http://www.greenseal.org/index.cfm; Green Electronics Council, "Welcome to GEC," 2009, http://www.greenelectronicscouncil.org.

[37] CRS Report R40850, *Managing Electronic Waste: Issues with Exporting E- Waste*, by Linda Luther.

[38] See, for example, Northwest Product Stewardship Council, "Electronic Equipment and Product Stewardship," June 29, 2009, http://www.productstewardship.net/productsElectronicsNWStates.html.

[39] Another term often used in the context of green procurement is *environmental attributes*. However, the term is used somewhat differently in different contexts. For example, in some cases it refers to a characteristic that is considered beneficial to the environment (see Federal Trade Commission, "Part 260 - Guides for the Use of Environmental Marketing Claims," April 12, 2007, http://www.ftc.gov/bcp/grnrule/guides980427.htm, C.F.R. Part 260). In other cases it may refer to a specific list of characteristics such as the potential for skin irritation or air pollution by a chemical (see Environmental Protection Agency, "Green Cleaning - Environmental Attributes Defined," January 28, 2010, http://www.epa.gov/epp/pubs/products/cleanattribut.htm).

[40] For example, a White House document from 2001 categorizes attributes as relating to natural resources use, human health and ecological stressors, hazardous factors associated with materials, and "positive attributes" such as recyclability (White House Task Force on Recycling, *Greening the Government: A Guide to Implementing Executive Order 13101*, February 2001). A European Commission study lists the following impacts: energy, land, resource, water, eutrophication, greenhouse gas, smog, acidification, and waste (European Commission, *Environmental Impact of Products (EIPRO)*, EUR 22284 EN, Technical Report Series, May 2006, http://ec.europa.eu/environment eipro_report.pdf).

[41] Commonly expressed as a percentage, AFUE "measures the amount of fuel converted to space heat in proportion to the amount of fuel entering the furnace" (Environmental Protection Agency and Department of Energy, "Furnaces Key Product Criteria," 2009, http://www.energystar.gov/index.cfm?c=furnaces.pr_crit_furnaces).

[42] SEER is the ratio of the cooling power of equipment and the electricity the equipment consumes (Environmental Protection Agency and Department of Energy, "What is SEER? EER? HSPF?," 2009, http://energystar.custhelp.com/ cgi-bin/energystar.cfg/php/enduser/std_adp.php?p_faqid=3041).

[43] Exceptions include green building (see CRS Report R40 147, *Issues in Green Building and the Federal Response: An Introduction*) and environmentally preferable purchasing programs (see footnote 35 and the text box on EPA's Environmentally Preferable Purchasing Program (EPP)).

[44] For example, one inventory program describes embodied energy as "the total primary energy consumed during resource extraction, transportation, manufacturing and fabrication of a product" (Sustainable Energy Research Team, University of Bath, "Embodied Energy & Carbon," 2007, http://www.bath.ac.uk/mech-eng/sert/embodied). Note that this definition does not include either use energy or energy consumed at end-of-life processes.

[45] Department of Energy, *2008 Buildings Energy Data Book*, November 2008, p. 1-32, http://buildingsdatabook.eren.doe.gov/docs%5CDataBooks%5CSEP_2008_BEDB.pdf. This meaning has been in use since at least the 1970s (see, for example, Committee on Science and Technology, Subcommittee on Energy Research, Development, and Demonstration, "Energy Accounting as a Policy Analysis Tool," committee print, 94th Cong., 2nd sess. (Washington: GPO, 1976).

[46] See, for example, Department for Environment, Food, and Rural Affairs, UK, *Progress Report on Sustainable Products and Materials*, July 2008, p. 24, http://www.defra.gov.uk/environment/

[47] Department of Energy, *2008 Buildings Energy Data Book*, Tables 1.6.2 and 1.6.3 Carbon emissions also tend to be higher for concrete.

[48] Methods have been developed for taking such variables into account in the field of energy analysis, but it is a very complex discipline, and its usefulness for procurement policy is not clear. For an overview, see Cutler J. Cleveland and Robert Costanza, "Net energy analysis," August 22, 2008, http://www.www.eoearth.org/article/Net_energy_analysis.

[49] See, for example, CRS Report RL34580, *Drought in the United States: Causes and Issues for Congress*, by Peter Folger, Betsy A. Cody, and Nicole T. Carter.

[50] A.K. Chapagain and A.Y. Hoekstra, *Water Footprints of Nations*, Value of Water Research Report Series No. 16 (UNESCO-IHE, November 2004), p. 42, http://www.waterfootprint.org/Reports/Report16Vol1.pdf. The estimates are based on a measure of water use called virtual water content, which is the volume of water used to produce a product (including production of crops for products with biobased content). The estimates given in the report were 8,000 liters of water for a pair of leather shoes, 10 liters for one sheet of letter-sized (A4) paper (one ream is 500 sheets), and 32 liters for one microchip.

[51] Environmental Protection Agency, "WaterSense," October 8, 2009, http://www.epa.gov/OW-OWM.html/water-efficiency/index.htm.

[52] Examples include Austin, Texas (http://www.ci.austin.tx.us/watercon); the Southern California Water District (http://socalwatersmart.com); and Seattle, Washington (http://www.savingwater.org/business_WSTP.htm).

[53] Water Footprint Network, "Water Footprint," 2010, http://www.waterfootprint.org/?page=files/home. An attempt to develop an international standard has also been initiated (see Sebastien Humbert, "ISO Standard on Water Footprint: Principles, Requirements and Guidance" (presented at the UNEP Water Accounting and Efficiency Stocktaking Workshop, Paris, France, November 23, 2009), http://www.unep.fr/scp/water ISO%20Activities%20on%20Water%20Accounting%20%28Sebastien%20Humbert%29.pdf).

[54] See, for example, National Research Council, *Perspectives on Biodiversity: Valuing Its Role in an Everchanging World* (Washington, DC: National Academy Press, 1999): "Existence values are values that are not predicated on use, in the ordinary sense of that word; that is, people gain utility from, or have preferences concerning, states of the world" (p. 84). This kind of value is difficult to measure and may be somewhat controversial, as are other kinds of so-called nonmarket values measured most commonly by contingent valuation techniques.

[55] This is briefly discussed in Environmental Protection Agency, "Final Guidance on Environmentally Preferable Purchasing for Executive Agencies," 64 Fed. Reg. 45810 (August 20, 1999). More detailed discussion can be found in White House Task Force on Recycling, *Greening the Government*. This guide refers specifically to several ecosystem impacts in the context of green procurement: endangered species, wetlands loss, fragile ecosystems, erosion, and animal welfare.

[56] Asbestos is present in many older buildings and is still used in some construction materials (Environmental Protection Agency, "Asbestos in Products and Buildings," December 10, 2008, http://www.epa.gov/asbestos pubs.html).

[57] Environmental Protection Agency, "Green Indoor Environments," October 14, 2008, http://www.epa.gov/iaq/greenbuilding.

[58] Environmental Protection Agency, *The National Biennial RCRA Hazardous Waste Report (Based on 2007 Data)*, EPA 530-R-08-012, November 2008, http://www.epa.gov/osw/inforesources/data/br07/national07.pdf.

[59] Ibid.

[60] The term *cost* as used here includes more than just purchase cost (see the section on "Cost").

[61] One of EPA's "guiding principles" for environmentally preferable purchasing is titled "Environment + Price + Performance = Environmentally Preferable Purchasing" (Environmental Protection Agency, "Final Guidance.").

[62] The proposed cleaner was much more concentrated than the one then in use and had lower preparation and disposal costs, for a total projected annual cost of $2,874, compared to $30,527 for the alternative. See Science Applications International Corporation, *Environmentally Preferable Purchasing* (The New York City Department of Sanitation, April 2001), p. 7, http://www.nyc.gov/html/nycwasteless/downloads/pdf/eppmanual.pdf.

[63] See, for example, City of Portland, Oregon, "Buying Green Case Studies," 2009, http://www.portlandonline.com/omf/?c=44701.
[64] Responsible Purchasing Network, *Responsible Purchasing Trends*.
[65] Generally, the term *life cycle assessment* is used to refer only to this kind of environmental assessment and should not be confused with similar terms with different meanings, such as life cycle management.
[66] See International Standards Organization, "13.020. 1 0:Environmental management," http://www.iso.org/iso/iso_catalogue/catalogue_ics/catalogue_ics_browse.htm?ICS1=13&ICS2=20&ICS3=10. See also Environmental Protection Agency, "LCA 101," October 17, 2008, http://www.epa.gov/ord/NRMRL/lcaccess/lca101.html.
[67] Leonardo Academy, "Development of the SCS-002 Life-Cycle Impact Declarations Standard," December 10, 2009, http://www.leonardoacademy.org/programs/standards/life-cycle/development.html.
[68] Science Applications International Corporation, *Life Cycle Assessment: Principles and Practice*, EPA/600/R-06/060 (Environmental Protection Agency, May 2006), http://www.epa.gov/ord/NRMRL/lcaccess/pdfs/600r06060.pdf.
[69] National Institute of Standards and Technology, "BEES 4.0," August 20, 2007, http://www.bfrl.nist.gov/oae/software/bees.
[70] The institute states that the tool uses data developed by the institute and data from the U.S. Life-Cycle Inventory (LCI) database developed by the Department of Energy (see Athena Sustainable Materials Institute, "EcoCalculator," April 20, 2009, http://www.athenasmi.org/tools/ecoCalculator/index.html; National Renewable Energy Laboratory, Department of Energy, "U.S. Life-Cycle Inventory Database," 2010, http://www.nrel.gov/lci/).
[71] See MBDC, *Cradle to Cradle Certification Program, Version 2.1.1*, September 2008, http://www.mbdc.com/docs/ Outline_CertificationV2_1_1.pdf, and **Table 3**.
[72] See, for example, Commission for Environmental Cooperation, "Green Building in North America," 2008, http://www.cec.org/greenbuilding.
[73] See, for example, Laurie Ramroth, *Comparison of life-cycle analyses of compact fluorescent and incandescent lamps based on rated life of compact fluorescent lamp*, Report No. C08-12 (Rocky Mountain Institute, February 2008), http://www.rmi.org/cms/Download.aspx?id=1393&file=C08-02_CFL_LCA.pdf.
[74] For example, a comparison of concrete and wood reveals much higher estimates for concrete of measures of potential impacts on energy, ecosystems, and health (Athena Institute, "EcoCalculator"). However, concrete may have lower impacts by other measures, depending for example on the degree to which companies use green practices in creating and using those materials.
[75] Using such a complex approach would involve fairly sophisticated mathematical modeling of effects across the life cycle, as well as information about the various elements, and as a result is unlikely to be feasible or practical for most procurement decisionmakers. Such analysis may be done for or used in technical studies (see, for example, Department for Environment, Food, and Rural Affairs, UK, *Progress Report*) or with the aid of tools such as BEES, but it is unlikely to be of direct use to many purchasers, such as small businesses. However, it may be particularly useful in the design stage, especially by manufacturers, and large purchasers such as the federal government in developing specifications for custom-made goods and services. Also, tools that may be developed to reduce the resources required to perform any given assessment may facilitate broader use. Finally, third-party certifying organizations may perform such analyses to determine whether a product or service should receive certification.
[76] A number of governmental and nongovernmental initiatives relating to such attributes are discussed in this chapter. With respect to corporate initiatives, an often-cited example is the efforts the Coca-Cola Company has made with respect to the water footprints of their products. See, for example, The Coca-Cola Company, *Replenish Report*, January 20 10, http://www.thecoca-colacompany.com/citizenship
[77] Christopher Robey, "Beyond Compliance: Environmental Sustainability in Federal Green Purchasing," *Contract Management* 49, no. 5 (May 2009): p. 20.
[78] For further discussion of these and other approaches, see, for example, Jerrell Coggburn and Dianne Rahm, "Environmentally Preferable Purchasing: Who is Doing What in the United States," *Journal of Public Procurement* 5, no. 1 (2005): 23-53; Case, *Environmental Purchasing Policies 10*.
[79] Environmental Protection Agency and Department of Energy, "Energy Star," 2009, http://www.energystar.gov. See also the Energy Star text box above.
[80] Green Electronics Council, "Welcome to EPEAT."
[81] Green Seal, "Green Seal."
[82] MBDC, *Cradle to Cradle Certification Program, Version 2.1.1*.
[83] Big Room, Inc., "Ecolabelling.org: Who's deciding what's green?," 2010, http://ecolabelling.org/.
[84] TerraChoice Environmental Marketing, *The Seven Sins of Greenwashing: Environmental Claims in Consumer Markets*, April 2009, http://sinsofgreenwashing.org/?dl_id=4.
[85] ISO has developed standards for ecolabeling that are used, for example, by Green Seal (International Organization for Standardization, "ISO 14020:2000 - Environmental Labels and Declarations—General Principles," 2009, http://www.iso.org/iso/iso_catalogue/catalogue_tc/catalogue_detail.htm?csnumber=34425).

Green Seal is also accredited by ANSI to develop standards, and some Green Seal standards are also ANSI standards (Green Seal, "How Green Seal Sets Environmental Standards," 2010, http://www.greenseal.org/certification

[86] The term has also been defined more broadly as "the act of misleading consumers regarding the environmental practices of a company or the environmental benefits of a product or service." According to one study, about half of the environmental claims for products that were for sale in various stores were vague or could not be proven (TerraChoice Environmental Marketing, *The Seven Sins of Greenwashing*).

[87] This may be fairly common, according to one study (ibid.).

[88] Responsible Purchasing Network, *Responsible Purchasing Trends*.

[89] But see the section on "Barriers to Broader Adoption of Green Procurement Practices" below.

[90] See the sections on "Cost" and "Measuring Success."

[91] This approach may also be called a *price preference* by some organizations, such as state and municipal governments (Case, *Environmental Purchasing Policies 101*).

[92] For example, Cincinnati, Ohio, established a 3% price preference for products containing recycled materials in 1994, and King County, Washington, established a price preference of up to 15% for recycled paper products in 1995 (ibid., pp. 11-13). Price preferences or premiums are not mentioned in the sections of the Federal Acquisition Regulations relating to green procurement (48 C.F.R. Part 23).

[93] Responsible Purchasing Network, *Responsible Purchasing Trends*.

[94] Examples include Oregon and Massachusetts (Case, *Environmental Purchasing Policies 101*, pp. 14-15). The Federal Acquisition Regulations include best value as part of the "Vision" ("All participants in the [Federal Acquisition Regulations] System are responsible for making acquisition decisions that deliver the best value product or service to the customer. Best value must be viewed from a broad perspective and is achieved by balancing the many competing interests in the System. The result is a system which works better and costs less" (48 C.F.R. § 1.102-1(b)); however, its explicit use appears to be limited in practice.

[95] For example, they may specify a percentage of contract dollars or total number of acquisitions that are to be green products or services.

[96] See, for example, Environmental Protection Agency, "Promoting Green Purchasing: Tools and Resources to Quantify the Benefits of Environmentally Preferable Purchasing," October 12, 2006, http://www.epa.gov/epp/tools/ epp_metrics.pdf; European Commission, *Environmental Impact of Products (EIPRO)*; Federal Facilities Environmental Stewardship and Compliance Assistance Center, "Acquisition (formerly Buying Green)," October 29, 2009, http://www.fedcenter.gov/programs/buygreen.

[97] Government Accountability Office, *Federal Procurement: Better Guidance and Monitoring Needed to Assess Purchases of Environmentally Friendly Products*, GAO-01-430, June 2001, http://www.gao.gov/new.items/d01430.pdf. No more recent assessments appear to have been done.

[98] Responsible Purchasing Network, *Responsible Purchasing Trends*, p. 16. The percentage was even lower (18%) for state government programs (Responsible Purchasing Network, *Responsible Purchasing Trends 2009: The 'State' of Sustainable Procurement*, 2009, http://www.responsiblepurchasing.org/publications/trends2009_naspo.pdf).

[99] Green Electronics Council, *Environmental Benefits of 2008 EPEAT Purchasing*, September 2009, http://www.epeat.net/Docs/Report2008_FullReport_R5.pdf.

[100] National Institute of Governmental Purchasing, *NIGP Survey of Green Procurement Initiatives*, July 2001, http://www.nigp.org/eweb/resourcelib/private/research/green.pdf.

[101] TerraChoice Environmental Marketing, *2009 EcoMarkets Summary Report*, September 2009, http://www.terrachoice.com/files/2009%20EcoMarkets%20Summary%20Report%20-%20September%2018,%202009.pdf.

[102] City of Portland, Oregon, "Buying Green Case Studies."

[103] See also CRS Report R40670, *Energy Efficiency in Buildings: Critical Barriers and Congressional Policy*, by Paul W. Parfomak, Fred Sissine, and Eric A. Fischer; several of the barriers discussed here are similar to those analyzed in that report with respect to buildings.

[104] For example, the Federal Trade Commission has developed guidance on the use of environmental marketing claims, both generally and with respect to recycling, compostability, and other end-of-life management claims, as well as recycled content, source reduction, reusability of containers, and damage to atmospheric ozone (Federal Trade Commission, "Part 260 – Guides."

[105] Green products and services are often perceived as being more expensive or less well-performing than standard alternatives, although there are many examples where such trade-offs do not in fact occur and a green design may also be a good design for both price and performance. However, investment in the assessments required to implement green procurement provide a significant first-cost barrier, especially for smaller enterprises.

[106] For example, pollutants put into waste streams may have impacts that are either broadly felt across a large population (such as greenhouse gases) or may affect completely different populations (such as acid rain).

[107] A classic example is from tenant-landlord relationships, in which a landlord has little incentive to improve the energy efficiency of a building if the tenant pays for utilities, and the tenant has little incentive in the absence of a lease period that is long enough to permit recoupment of the investment.

[108] A number of federal regulatory programs, such as those relating to toxic substances, may result in greener products and services in the marketplace. However, since one effect of such regulations is to make such products and services industry standard, discussion of those programs is beyond the scope of this chapter, which focuses on practices that are greener than the standard ones. For a discussion of current legislative issues associated with such regulatory programs, see CRS "Issues in Focus" on *Energy, Environment, and Resources.*

[109] See 48 C.F.R. Subparts 23.2 and 23.4. Unlike for Energy Star, the BioPreferred program name is not mentioned in the regulation except as part of a URL to which contracting officers are referred (http://www.usda.gov/biopreferred).

[110] See the section on "Green Procurement and Environmentally Preferable Purchasing."

[111] For a discussion of federal market-focused programs relating to green building, see CRS Report R40147, *Issues in Green Building and the Federal Response: An Introduction.*

[112] Environmental Protection Agency, "Database of Environmental Information for Products and Services," April 2, 2010, http://yosemite1.epa.gov/oppt/eppstand2.nsf.

[113] Federal Trade Commission, "FTC Consumer Information - Energy & the Environment: Environment," August 11, 2009, http://www.ftc.gov/bcp/menus/consumer/energy

[114] Federal Trade Commission, "Part 260 - Guides."

[115] In addition to RCRA, they include the Clean Air Act Amendments of 1990 (P.L. 10 1-549), the Energy Policy Act of 1992 (EPACT 1992, P.L. 102-486), the Farm Security and Rural Investment Act (P.L. 107-17 1), the Energy Policy Act of 2005 (EPACT 2005, P.L. 109-58), and the Energy Independence and Security Act of 2007 (EISA, P.L. 110-140).

[116] See especially 48 C.F.R. Part 23.

[117] See especially Executive Order 13423, "Strengthening Federal Environmental, Energy, and Transportation Management," *Federal Register* 72, no. 17 (January 26, 2007): 3919-3923; and Executive Order 13514, pp. 52117- 52 127. For a discussion of the relationship between those two executive orders, see CRS Report R40974, *Executive Order 13514: Sustainability and Greenhouse Gas Emissions Reduction*, by Richard J. Campbell and Anthony Andrews.

[118] The term *policy documents* here means a federal agency document, other than an enacted statute, an executive order, or a regulation, that is used to set, implement, or interpret policies. Examples cited elsewhere in this chapter are Office of Management and Budget, "Instructions for Implementing Executive Order 13423," March 29, 2007, http://www.fedcenter.gov/_kd/Items/actions.cfm?action=Show&item_ id=6825&destination Department of Defense, "Department of Defense Green Procurement Program Strategy," November 2008, http://www.acq.osd.mil/ dpap/cpic/cp/docs/USA001967-08-DPAP.pdf.

[119] See footnote 39.

[120] For example, § 104 of EISA contains a requirement that "[i]n the case of electric motors of 1 to 500 horsepower, agencies shall select only premium efficient motors that meet a standard designated by the Secretary [of Energy]" (119 Stat. 610).

[121] See, for example, U.S. Department of Agriculture, "BioPreferred."

[122] This includes products listed by Energy Star and by the Federal Energy Management Program (FEMP; Department of Energy, "Federal Energy Management Program," October 21, 2009, http://www w1.eere.energy

[123] See text box on page 35.

[124] For regulations, see Part 23 of the Federal Acquisition Regulation (48 C.F.R. Part 23), available at https://www.acquisition.gov/far. Relevant statutes are listed in footnote 115.

[125] However, green building can also be considered as a special kind of green procurement that also involves a multiattribute, integrative approach. See CRS Report R40147, *Issues in Green Building and the Federal Response: An Introduction.*

[126] See Environmental Protection Agency, "Database of Environmental Information for Products and Services."

[127] Department of Energy, "Federal Energy Management Program: Energy Savings Performance Contracts," September 3, 2009, http://www1.eere.energy

[128] See, for example, 48 C.F.R. § 23.405, stating that contracting officers should refer to lists of EPA- and USDA-designated items of recovered-material and biobased products "when purchasing services that could include supplies that contain" such materials.

[129] See Office of Management and Budget, "Federal Procurement Policy - Green Acquisition," 2009, http://www.whitehouse.gov/omb/procurement

[130] Environmental Protection Agency, *Integrating Green Purchasing into Your Environmental Management System (EMS)*, EPA 742-R-05-001, April 2005, p. 38, http://www.epa.gov/epp/pubs/ems.htm.

[131] See, for example, Office of Management and Budget, "Instructions for Implementing Executive Order 13423."

[132] See, for example, Office of Management and Budget, "Policy Letter 92-4," November 2, 1992, http://www.whitehouse.gov/omb/procurement

[133] Office of Management and Budget, "FY 2008 Reporting on Green Purchasing Requirements," Memorandum for Federal Agency Environmental Executives and Agency Senior Procurement Executives, October 20, 2008, http://www.whitehouse.gov/omb/procurement

[134] Office of Management and Budget, "Instructions for Implementing Executive Order 13423."

[135] See Green Electronics Council, "Welcome to EPEAT."

[136] Environmental Protection Agency, "Database of Environmental Information for Products and Services."

[137] Environmental Protection Agency, "Significant New Alternatives Policy (SNAP) Program," October 19, 2009, http://www.epa.gov/Ozone/snap/index.html.

[138] Environmental Protection Agency, "WaterSense."

[139] Biobased products are commercial or industrial products or ingredients other than food or feed that are composed of biological materials and for which there is not a mature market. The focus is on renewable materials. See Department of Agriculture, "BioPreferred."

[140] General Services Administration, "Environmental Products Overview." State and local government agencies may also make purchases through GSA.

[141] General Services Administration, "Recycling Personal Property Disposal," July 10, 2009, http://www.gsa.gov/Portal/gsa/ep/channelView.do?pageTypeId=171 10&channelPage=%2Fep%2Fchannel%2FgsaOverview.jsp&channelId=-24484.

[142] Office of the Federal Environmental Executive, "Green Products Compilation," April 2, 2010, http://www.fedcenter.gov/Documents/index.cfm?id=11767&pge_prg_id=26960.

[143] See, for example, CRS Report R40168, *Alternative Fuels and Advanced Technology Vehicles: Issues in Congress*, by Brent D. Yacobucci.

[144] See, for example, House Subcommittee on Government Management, Organization, and Procurement, Committee on Oversight and Government Reform, *IT Procurement and Disposal: Application of the Federal Government's Green Policies in the Life Cycle Management of IT Assets.*, 2009, http://oversight task=view&id=4627&Itemid=28.

[145] CRS Report R40147, *Issues in Green Building and the Federal Response: An Introduction*.

[146] See, for example, Office of Management and Budget, "FY 2008 Reporting on Green Purchasing Requirements." The memorandum states in part,

Section 6002 of the Resource Conservation and Recovery Act (RCRA) and section 9002 of the Farm Security and Rural Investment Act (FSRIA) require the Office of Federal Procurement Policy (OFPP) to report to Congress every two years on the actions taken by federal agencies to implement the purchasing requirements of these statutes. E.O. 13423 (Order) requires the OMB Director to issue instructions concerning implementation of the Order's acquisition requirements and periodic evaluation of agencies' implementation of the Order. It also directs the Federal Environmental Executive to report to the President every two years on agencies' activities to implement the Order. Heads of agencies are directed to provide reports on agency implementation as required.

[147] In 2001, the Government Accountability Office reported that a White House task force had recommended including information on recycled-content purchases in the FPDS (Government Accountability Office, *Federal Procurement: Better Guidance and Monitoring Needed to Assess Purchases of Environmentally Friendly Products*).

[148] Percentages are estimates derived from reports generated by CRS using the FPDS, https://www.fpds.gov.

[149] 42 U.S.C. § 6374(a)(3)(E)(ii) requires the Secretary of Energy to monitor compliance of agency fleets with alternative-fuel vehicle requirements and to report annually to Congress on the level of compliance, including annual reductions in use of petroleum-based fuels and any problems in acquiring alternative fuels.

[150] Department of Energy, *Federal Fleet Compliance with EPA CT and E.O. 13423: Fiscal Year 2007*, 2009, http://www1.eere.energy

[151] Office of Federal Procurement Policy, Office of Management and Budget, *Report on Agency Implementation of Buy-Recycled and Buy-Biobased Requirements in the Resource Conservation and Recovery Act and the Farm Security and Rural Investment Act*, January 16, 2009, http://www.whitehouse.gov/omb/assets rcra_and_fsria_rpt.pdf.

[152] Office of Management and Budget, "FY 2008 Reporting on Green Purchasing Requirements."

[153] Department of Defense, "Department of Defense Green Procurement Program Strategy."

[154] Defense Logistics Agency, "Green Procurement Report (GPR)," October 1, 2009, http://www.dlis.dla.mil/erlsgpr/default.asp. In an FY2006 report to Congress, DOD identified attributes for which it tracked procurement as including recovered and recycled materials, energy efficient, low volatile organic compounds, water conserving, and asbestos alternatives (see Department of Defense, *Fiscal Year 2006 Defense Environmental Programs Annual Report to Congress*, 2007, Appendix W, https://www.denix.osd.mil/portal/page/portal/denix/environment

[155] Department of Energy, "U.S. Department of Energy Environmentally Preferable Purchasing Program Handbook," April 2008, http://www.hss.doe.gov/pp/epp/EPP-DOE-Handbook-rev12.pdf.

[156] Office of Federal Procurement Policy, *Report on Agency Implementation of Buy-Recycled and Buy-Biobased Requirements*.

[157] Juan D. Lopez, "OMB Scorecards: Energy, Transportation, and Environmental Management," January 30, 2007, http://www.federalelectronicschallenge.org/resources

[158] For summary scorecards, see Federal Facilities Environmental Stewardship and Compliance Assistance Center, "EO 13423," November 9, 2009, http://www.fedcenter.gov/_kd/go.cfm?destination

[159] For example, E.O. 13423 sets a policy goal that federal agencies "conduct their environmental, transportation, and energy-related activities under the law in support of their respective missions in an environmentally, economically and fiscally sound, integrated, continuously improving, efficient, and sustainable manner" (p. 3919), and OMB's implementing instructions state that in meeting this goal, "[e]ach agency shall consider life-cycle costs and savings in planning and making determinations about investments in all capital assets, services, and procurements...." (p. 4). E.O. 13514 sets the goal of the federal government leading by example in the creation of a "clean energy economy," and lists several relevant policies, including "leverag[ing] agency acquisitions to foster markets for sustainable technologies and environmentally preferable materials, products, and services" (p. 52117).

[160] For example, should green procurement be considered the integrated, continuous-improvement approach to reducing the environmental footprint of procured products and services described in the first part of this chapter? Is it a synonym for EPA's existing EPP program? Is it an umbrella term describing any procurement activity that takes environmental matters into account, no matter how limited?

[161] See footnote 118.

[162] See footnote 159.

[163] See, for example, Environmental Protection Agency, *Energy Star—the Power to Protect the Environment*; Green Electronics Council, *Environmental Benefits of 2008 EPEAT Purchasing*, September 2009, http://www.epeat.net/Docs/ Report2008_FullReport_R5.pdf.

[164] There are some GAO studies, but they have focused on specific initiatives or activities associated with green procurement. See, for example, Government Accountability Office, *Federal Electronics Management: Federal Agencies Could Improve Participation in EPA's Initiatives for Environmentally Preferable Electronic Products*, GAO- 10-196T, October 27, 2009, http://www.gao.gov/new.items/d10196t.pdf;———, *Federal Procurement: Better Guidance and Monitoring Needed to Assess Purchases of Environmentally Friendly Products*.

[165] Percentages vary from year to year and depend also on how the calculation is done. Data from Table 1 yield a result of 45% for FY2008 and 49% for FY2009, whereas a recent report cited a figure of 64% for FY2004 and 60% for FY2005 (See Acquisition Advisory Panel, Report of the Acquisition Advisory Panel to the Office of Federal Procurement Policy and the United States Congress, January 2007, p. 440, http://www.acquisition.gov/comp/aap/ 24102 _GSA.pdf; the figures are from a custom analysis performed for the panel by FPDS. The figures cited in that report may treat construction as a service; the percentage of total spending on services without construction for FY2005, calculated by CRS from FPDS-NG data, was 54%, with construction comprising an additional 6% of total spending.

[166] 42 U.S.C. § 6962(f).

[167] See 72 Fed. Reg. 73907-08.

[168] See Environmental Protection Agency, "Final Guidance."

[169] 48 C.F.R. § 23 .405(c).

[170] The FAR states only that it is government policy "to acquire supplies and services that promote ... water efficiency" (48 C.F.R. § 23 .202) and that agencies must "employ acquisition strategies that ... promote ... water conservation" (48 C.F.R. § 23.703(b)), but there are no more explicit requirements, unlike with energy efficiency and some other preferences.

[171] See, for example, "OFPP Calls for 5-Percent Increase in the Civilian Acquisition Workforce," 92 *Fed. Cont. Rep.* 311 (Nov. 3, 2009).

[172] *Cf.* Kellie Lunney, "Green Government," *Government Executive*, August 1, 2008, http://www.govexec.com/ story_page.cfm?articleid=40616.

Chapter 3

FEDERAL AGENCIES COULD IMPROVE PARTICIPATION IN EPA'S INITIATIVES FOR ENVIRONMENTALLY PREFERABLE ELECTRONIC PRODUCTS

John B. Stephenson

WHY GAO DID THIS STUDY

Advancing technology has led to increasing sales of new electronic devices. With this increase comes the dilemma of managing them at the end of their useful lives. If discarded with common trash, a number of environmental impacts may result, ranging from the loss of valuable resources to the potential release of toxic substances, such as lead. If recycled, they may be exported to countries with waste management systems that are less protective of human health and the environment that those of the United States.

The federal government is the world's largest purchaser of electronics, spending nearly $75 billion on electronic products and services in 2009. The Environmental Protection Agency (EPA) has helped implement several product stewardship initiatives to encourage responsible management of electronic products in all three phases of a product's lifecycle—procurement, operation, and end-of-life disposal. In response to a request to provide information on federal procurement and management of electronic products, GAO's testimony describes (1) EPA's electronic product stewardship initiatives, (2) federal agency participation in them, and (3) opportunities for strengthening participation. GAO's testimony is based on its prior work and updated with data from EPA. In our prior report, EPA agreed that increasing federal participation in its initiatives could be encouraged. Agency officials still agree with this finding.

WHAT GAO FOUND

Federal government approaches to ensuring environmentally responsible management of electronic equipment from procurement through disposal rely heavily on two interrelated initiatives. The first initiative, the electronic product environmental assessment tool (EPEAT®), was developed along the lines of EPA's and the Department of Energy's Energy Star program and assists federal procurement officials in comparing and selecting computers and monitors with environmental attributes that also routinely save money through reduced energy usage over the products' lives. The second initiative— the federal electronics challenge (FEC)—helps federal agencies realize the benefits of EPEAT-rated electronics by providing resources to help agencies extend these products' life spans, operate them in an energy efficient way, and expand markets for recovered materials by recycling them at end of life.

The first 5 years of EPA's initiatives have resulted in notable energy savings and environmental benefits reported by participating agencies. According to facilities that reported information to EPA and the Office of the Federal Environmental Executive in 2008, 88 percent of all desktop computers, laptop computers, and monitors the facilities purchased or leased were EPEATregistered. EPEAT participation reportedly resulted in procurement officials purchasing 95 percent of their monitors with Energy Star power management features enabled and 38 percent of computers with this feature. In addition, 16 federal agencies and 215 federal facilities—representing about one-third of all federal employees—participated in the FEC to some extent in 2008. As a result, participants reported that 50 percent of electronics taken out of service were donated for reuse, 40 percent were recycled, 8 percent were sold, and 2 percent were disposed of. The environmentally responsible choices associated with EPEAT and FEC resulted in a reported $40.3 million in cost savings for participants.

The EPEAT and FEC accomplishments are steps in the right direction, but opportunities exist to increase the breadth and depth of federal participation. First, agencies and facilities representing about two-thirds of the federal workforce are not participating in these promising initiatives, despite instructions to do so in implementing Executive Order 13423. Second, few participating agencies and facilities maximize these programs' resources and their potential benefits. For some, participation simply means the agency identified its current practices for managing electronic products and set goals to improve them. Moreover, as the FEC aims to support participating agencies and facilities, it does not impose consequences for those that do not meet their goals. In fact, only 34 FEC facility partners showed they managed electronic products in 2008 in accordance with FEC goals for at least one of the three lifecycle phases, and only 2 facilities showed they did so for all phases. For perspective, GAO calculated that if federal agencies replaced 500,000 desktop and laptop computers and monitors with EPEAT-registered products and operated and disposed of them in accordance with FEC goals, they could achieve substantially greater energy reductions and cost savings.

Chairwoman Watson and Members of the Subcommittee:
I am pleased to be here today to discuss findings from our work on federal procurement of environmentally preferable electronic products and ways in which such procurement can lessen the impacts of electronic waste (e- waste) disposal. The federal government is the

world's largest purchaser of information technology equipment, annually spending nearly $75 billion on electronic products and services. Through its purchasing decisions, the federal government has substantial leverage to enhance recycling infrastructures and stimulate markets for environmentally preferable electronic products. Along these lines, the Environmental Protection Agency (EPA) has helped implement several product stewardship initiatives under its Resource Conservation Challenge. These initiatives encourage environmentally responsible management of electronic products from "cradle to grave"—that is, from the initial procurement of environmentally preferable products, to their operation in an energy efficient manner, and finally to their reuse or recycling in an environmentally safe way.

Disposing of e-waste has become an important issue as rapidly advancing technology has led to increasing sales of new electronic products—in particular, computers, monitors, and handheld devices such as cell phones. With this increase comes the dilemma of managing these products at the end of their useful lives. Little information exists, for example, on whether obsolete electronic products are reused, stored, or disposed of in landfills. As we previously reported, if discarded with common trash, a number of adverse environmental impacts may result, ranging from the loss of valuable resources in the electronics such as copper, gold, and aluminum to the potential for harmful substances such as cadmium, lead, and mercury to enter the environment.[1] If donated or recycled, these products may eventually be irresponsibly exported to countries without modern landfills and with waste management systems that are less protective of human health and the environment than those in the United States. In our August 2008 report, we showed that e-waste exported from the United States to developing countries, such as those in Southeast Asia, is often dismantled under dangerous health conditions, using methods like open-air incineration and acid baths to extract precious metals.[2]

Our testimony, which is based on our prior work and updated with data from EPA,[3] provides observations on (1) EPA's electronic product stewardship initiatives, (2) the extent of federal agency participation in them, and (3) opportunities for strengthening participation. Our prior work was conducted in accordance with generally accepted government auditing standards. Those standards require that we plan and perform the audit to obtain sufficient, appropriate evidence to provide a reasonable basis for our findings and conclusions based on our audit objectives. We believe that the evidence obtained provides a reasonable basis for our findings and conclusions based on our audit objectives.

BACKGROUND

The purchase price of electronic products primarily reflects their technological capabilities; it does not include all of the substantial costs that are incurred throughout the equipment's life. A study by Gartner Research, for example, shows that computers costing less than $1,000 typically have a total cost of ownership of more than $5,000 per year when all the energy and maintenance costs are included.[4] Furthermore, the purchase price of electronics does not include the often substantial cost of disposal. Lifecycle costs are high, in part, because electronic products are not always designed to facilitate recycling.

EPA estimates that across the federal government 10,000 computers are disposed of each week. Once such products reach the end of their original useful lives, federal agencies have several options for disposing of them. Agencies generally can donate their reusable equipment to schools or other nonprofit educational institutions; give them to a recycler; exchange them with other federal, state, or local agencies; sometimes trade them with vendors to offset the costs of new products; or sell them through the General Services Administration's (GSA) surplus property program, which sells surplus federal government equipment at public auctions.

Federal agencies, however, are not required to track the ultimate destination of their donated or recycled e-waste. Instead, agency officials generally consider this to be the recipient organization's responsibility. Consequently, they often have little assurance that their e-waste is ultimately disposed of in an environmentally responsible manner. In our prior work, we found that some U.S. electronics recyclers—including ones that publicly tout their exemplary environmental practices—were apparently willing to circumvent U.S. hazardous waste export laws and export e-waste to developing countries. Specifically, we posed as foreign buyers of broken cathode-ray tube computer monitors—which are considered hazardous waste and illegal to export without a permit—in Hong Kong, India, Pakistan, and other countries; and 43 U.S. companies expressed willingness to export these items. Some of the companies were willing to export this equipment in apparent violation of U.S. law. As we showed in our August 2008 report,[5] equipment exported to developing countries may be handled in a way that threatens human health and the environment.

TWO PROMISING INITIATIVES ASSIST FEDERAL AGENCIES IN PROCURING, OPERATING, AND DISPOSING OF ELECTRONIC PRODUCTS IN AN ENVIRONMENTALLY PREFERABLE MANNER

As we reported in November 2005,[6] existing federal government approaches to ensuring environmentally responsible management of electronic equipment from procurement through disposal rely heavily on two interrelated EPA electronic product stewardship initiatives. The first, the electronic product environmental assessment tool (EPEAT®), assists federal procurement officials in comparing and selecting laptop computers, desktop computers, and monitors with environmentally preferable attributes. The second, the federal electronics challenge (FEC), helps federal agencies fully utilize the benefits of EPEAT-rated electronics by providing resources to help agencies extend these products' life spans, operate them in an energy efficient way, and expand markets for recycling and recovered materials by recycling them at end of life.[7]

EPEAT was developed along the lines of EPA's and the Department of Energy's (DOE) Energy Star program in which the federal government rewards manufacturers of energy-efficient products that ultimately save money and protect the environment by providing them with a label for their products that certifies these benefits. EPEAT-registered products are awarded a bronze, silver, or gold certification for increasing levels of energy efficiency and environmental performance. Using EPEAT, an online tool, federal procurement officials can evaluate the design of an electronic product for energy conservation, reduced toxicity, extended lifespan, and end of life recycling, among other things. For example, EPEAT can help agency procurement officials choose electronic products with attributes that make the products easier to upgrade. Some computers are now being built with modular features so that hard drives, processors, memory cards, and other components can be upgraded rather than replaced—thus extending their lifecycles. Agency procurement officials can also use EPEAT to choose among products that are designed to make recycling less expensive, such as those without glues or adhesives, with common fasteners and "snap-in" features, and with easily separable plastic and metal components—making their disassembly easier and recycling less costly. Finally, EPEAT can help procurement officials identify electronic products that contain less hazardous materials, which can also lessen their disposal and recycling costs.

Products with these attributes can, in many cases, save agencies money over the products' lifecycles when compared to those with similar technological characteristics but without environmental attributes. For example, according to one computer vendor, a particular desktop computer with energy-saving attributes cost $35 more than a similar model that one federal program office had been buying; however, it will save $15 per year in energy costs. Thus, after slightly more than 2 years of use, the EPEAT-rated desktop computer can save more money in energy savings than the additional increase in purchase price and result in measurable environmental benefits.

Currently, in the electronic products industry, purchasers can choose from 170 desktop computers, 637 laptop computers, and 487 monitors that meet one of the three EPEAT levels of environmental performance. The breadth of EPEAT products provides procurement officials with a range of devices to meet their technology and budgetary needs. For example, agencies have the flexibility to choose liquid crystal display monitors that meet all the required EPEAT criteria as well as numerous optional criteria, such as the lower levels of mercury in light switches and a reduced number of different types of plastics—attributes that can make recycling easier and less costly. Agencies can also choose other monitors that meet these and other criteria, including additional reductions in toxic materials, along with end-of-life services such as a take-back and reuse program for packaging material. Of note, these different types of monitors can meet different technology needs, as there are some differences in display characteristics and power consumption.

As we said earlier, federal agencies also have the opportunity to participate in FEC—a program that first relies heavily on EPEAT for procurement considerations and then provides guidance to participants on how to extend electronic product life spans, operate them in an energy- efficient way, and reuse or recycle them at end-of-life. FEC differs from EPEAT in that where EPEAT assists officials in procuring environmentally preferable products, FEC provides participating agencies and facilities with resources to help ensure that electronic products are operated and disposed of in a manner that fully utilizes the environmental attributes of the EPEAT product.[8] FEC has two partner levels: agency and facility. To participate, executive branch agencies or their subcomponents must register.

According to EPA documents, participation can provide agency officials greater assurance that the e-waste they donate to schools, or send for recycling, is ultimately disposed of in an environmentally responsible manner.[9] For instance, in following FEC guidance, participants are to provide recipients of donated equipment with instructions on how to have the equipment recycled responsibly and how to verify that responsible recycling occurs—procedures known as "downstream auditing." When donating equipment, FEC instructs agencies and facilities to ensure that recipients contact local or state environmental or solid waste agencies to obtain a database of vendors who recycle e-waste once the equipment is no longer useful to the recipient organization.

FEC also recommends that participating agencies and facilities instruct recipients to avoid arrangements with recyclers that are unable or unwilling to share references and cannot explain the final destination of the e-waste they collect. When recycling equipment, participants are to determine how much electronic equipment the recyclers actually recycle, versus the amount they sell to other parties. If the majority of the incoming e-waste is sold, the recycling facility may be sending a significant amount of e-waste into landfills or for export overseas. In addition, FEC instructs participants to physically inspect potential recycler's facilities. E-waste in trash containers, for example, may indicate that the facility is not recycling it, and the presence of shipping containers may indicate that the facility exports it.

FEDERAL AGENCIES AND FACILITIES HAVE INCREASED PARTICIPATION IN EPEAT AND FEC IN RECENT YEARS

As of December 31, 2008, EPA reported that 16 federal agencies and 215 federal facilities—representing slightly more than one-third of all federal employees—participated in the FEC to some extent. In addition, according to the 128 facilities that reported data to EPA, a majority of electronic products purchased during 2008 were EPEAT-registered. This is a sizeable increase from 2005, when we reported that 12 federal agencies and 61 individual federal facilities participated in FEC. Participating agencies include the Departments of Agriculture, Commerce, Defense, Energy, Health and Human Services, Homeland Security, Interior, Justice, Labor, Treasury, Transportation, and Veterans Affairs, as well as the Environmental Protection Agency, Executive Office of the President, General Services Administration, and the United States Postal Service.[10]

The benefits of federal agency and facility participation in EPEAT and FEC offer a glimpse of what can be attained through greater federal involvement. For instance, in 2008 FEC participants reported to EPA and the Office of the Federal Environmental Executive that 88 percent of all desktop computers, laptop computers, and monitors they purchased or leased were EPEAT registered. In addition, FEC participants reported that they extended computer life spans so that 63 percent of computers had at least a 4-year useful life. Procurement officials reported purchasing 95 percent of their monitors with energy-efficient power management features enabled and 38 percent of computers with this feature. Finally, participants reported that 50 percent of electronics taken out of service were donated for reuse; 40 percent were recycled; 8 percent were sold; and 2 percent were disposed of. Of those recycled, 95 percent were reportedly done so in an environmentally sound manner. These environmentally preferable choices from "cradle to grave" resulted in $40.3 million in cost savings reported by participating agencies and facilities, energy savings that EPA found to be equivalent to electric power for more than 35,000 U.S. households for 1 year, and emissions savings equivalent to removing nearly 21,000 passenger cars from the road for 1 year.[11]

Table 1. Environmental Benefits of Agencies Procuring Computers and Monitors that Meet EPEAT's Bronze, Silver, or Gold Level of Environmental Performance and Operating and Disposing of Them in Accordance with FEC Goals

	Reduction in Energy Usage (kWh)	Reduction in Raw Material Extraction (kg)	Reduction in Greenhouse Gas Emissions (kg)	Reduction in Toxic Materials (kg)
EPEAT-Bronze				
Procurement	383,000,000	685,000,000	72,800,000	41,500
Operation	1,010,000,000	1,750,000,000	192,000,000	2,540
Disposal	794,000,000	10,800,000	42,400,000	8,310
Bronze Total	**2,187,000,000**	**2,445,800,000**	**307,200,000**	**52,350**
EPEAT-Silver				
Procurement	388,000,000	686,000,000	73,000,000	41,500
Operation	1,010,000,000	1,750,000,000	192,000,000	2,540
Disposal	794,000,000	10,800,000	42,400,000	8,310
Silver Total	**2,192,000,000**	**2,446,800,000**	**307,400,000**	**52,350**
EPEAT-Gold				
Procurement	393,000,000	687,000,000	73,300,000	41,500
Operation	1,010,000,000	1,750,000,000	192,000,000	2,540
Disposal	794,000,000	10,800,000	42,400,000	8,310
Gold Total	**2,197,000,000**	**2,447,800,000**	**307,700,000**	**52,350**

Source: EPA environmental benefits calculator.

Through participation in the FEC, numerous federal facilities have purchased greener electronic products, reduced the environmental impacts of electronic products during use, and managed obsolete electronics in an environmentally safe way. For example, officials with the Bonneville Power Administration within DOE reported to EPA that they adopted several environmentally responsible practices associated with the procurement and operation of electronic equipment. First, administration officials extended the lifespan of agency computers from 3 to 4 years. With over 500 computers procured each year at an annual cost of more than $500,000, an administration official said that extending computer life spans generated substantial savings. Additionally, Bonneville Power Administration officials procured new flat-screen monitors instead of cathode-ray tube monitors, reducing both hazardous waste tonnage and end of life recycling costs. According to Bonneville Power Administration officials, they expect to save at least $153 per unit over the life of each new monitor.

EPA's region 9 facility in San Francisco, California—a 20-story office building that houses nearly 900 EPA employees—also reported achieving substantial environmental benefits through participation in the FEC. The facility's energy subcommittee recommended an audit, which found that enabling computer and monitor power management features, such as those configuring computer monitors to the "sleep" mode instead of the screen saver mode, could save about 10 percent in total energy usage at no cost. In addition, with funding eliminated for new electronics purchases, region 9 staff reported that they reused 30 percent to 40 percent of existing electronics and extended the average lifespan of computers to 5 years. Finally, region 9 staff stated that they successfully recycled more than 10 tons of electronics that had been stored in an offsite warehouse. Although the cost of safely recycling the large quantity of electronics was high and regional staff found it difficult to locate a reputable recycler, EPA headquarters provided funds for the recycling costs and helped find a qualified vendor.

OPPORTUNITIES EXIST FOR MORE FEDERAL AGENCIES AND FACILITIES TO JOIN EPA'S INITIATIVES, AND CURRENT PARTICIPANTS CAN SIGNIFICANTLY STRENGTHEN THEIR PARTICIPATION

The EPEAT and FEC accomplishments achieved to date are steps in the right direction, but opportunities exist to significantly increase the breadth and depth of federal agency and facility participation. First, agencies and facilities representing almost two-thirds of the federal workforce are not yet participating in these promising initiatives, despite Executive Order 13423.[12] This executive order, signed by the President on January 24, 2007, generally requires that each agency (1) meets at least 95 percent of its requirements with EPEAT-registered products; (2) enables the energy saving features on agency computers and monitors; (3) establishes and implements policies to extend the useful life of agency electronic equipment; and (4) uses environmentally sound practices with respect to disposition of agency electronic equipment that has reached the end of its useful life. To implement these requirements, the Office of Management and Budget directed each agency and its facilities to either become a partner in the FEC or to implement an equivalent electronics stewardship program that addresses purchase, operation and maintenance, and end-oflife management strategies for electronic assets consistent with FEC's recommended practices and guidelines.

Second, most of agencies and facilities that participate do not fully maximize these programs' resources or the environmental benefits that can be achieved. While we acknowledge the efforts of FEC participants, the FEC statistics on participation may overstate these participants' adherence to the goals of the program, and their successes must be taken in context. Participation by 16 agencies and 215 facilities (representing slightly more than one-third of federal employees), for example, does not mean that all electronic products they purchase are procured, operated, and recycled or reused at end of life in an environmentally preferable fashion. Instead, participation simply means these agencies have identified their current practices for managing electronic products and set goals to improve them. Moreover, as the FEC is an initiative aimed to encourage and support participating agencies and facilities, it does not impose consequences on those agencies who do not meet their goals. As a practical matter, only 34 FEC facility participants (16 percent of participants) reported to EPA that they managed electronic products in accordance with FEC goals for at least one of the three lifecycle phases— procurement, operation, or disposal—with only 2 facilities showing they did this for all three phases in 2008.[13]

The need for increased federal participation in these initiatives—in both breadth and depth—is further underscored by the federal government e- waste that continues to appear in online auctions and may subsequently end up overseas. As we reported in August 2008,[14] significant demand exists for used electronics from the United States. We observed thousands of requests for such items on e-commerce Web sites—mostly from Asian countries, such as China and India, but also from some African countries. In our prior work, we showed that these countries often lack the capacity to safely handle and dispose of e-waste, as disassembly practices in these countries often involve the open-air burning of wire to recover copper and open acid baths for separating metals. These practices expose people to lead and other hazardous materials. In the several weeks leading up to this hearing, we monitored an e-commerce Web site where surplus federal government equipment is auctioned and found nearly 450,000 pounds of cathode-ray tube monitors for sale—items that, based on our prior work, have a high likelihood of being exported.

For perspective, using EPA's environmental benefits calculator[15] we calculated the benefits that would result under a hypothetical scenario in which federal agencies replaced 500,000 desktop and laptop computers and computer monitors using EPEAT procurement criteria for each tier of environmental performance—bronze, silver, and gold. As part of this calculation, we added the environmental benefits attained if federal agencies operated all EPEAT units in an energy efficient manner (i.e., enabled Energy Star features) and reused and recycled the end-of-life electronics they replaced in accordance with FEC goals. We found that substantial energy savings and environmental benefits would result at all three EPEAT tiers. Specifically, greater participation could lead to environmental benefits 5- to 10-times greater than the accomplishments of FEC participants in 2008 described earlier. Additionally, if federal agencies were to purchase EPEAT-bronze, silver, or gold products, according to the EPA environmental benefits calculator, they would save approximately $207 million at each level of EPEAT performance in energy usage and realize other cost, waste, and emissions reductions over the useful lives of these products. Table 1 shows the net energy savings and reductions in raw material extraction, greenhouse gas emissions, and toxic materials that would result if agencies and facilities recycled electronic products and replaced them with EPEAT-rated units, as compared to non-EPEAT computers and monitors.[16]

To help agency officials put in context the environmental and economic benefits that can result from using environmentally preferable electronic products, the EPA environmental benefits calculator also shows the benefits of procurement, operation, and disposal in accordance with FEC goals using common equivalents. Table 2 shows the environmental benefits of these practices when measured as the amount of household energy usage saved annually and the volume of automobile emissions saved annually.

Table 2. Common Equivalents to the Environmental Benefits of Procuring, Operating, and Disposing of Computers and Monitors in Accordance with FEC Goals

	Number of U.S. Households' Energy Usage Saved	Number of Passenger Cars Off Roadways
EPEAT-Bronze	182,796	206,257
EPEAT-Silver	183,151	206,349
EPEAT-Gold	183,570	206,543

Source: EPA environmental benefits calculator

CONCLUDING OBSERVATIONS

Understandably, when procuring electronics in a challenging fiscal environment, agency officials may give greater weight to price than environmental attributes. However, many of the environmental and human health problems associated with e-waste disposal can be averted through environmentally preferable procurement. Using EPEAT to purchase environmentally-friendly products, agency purchasers can often simultaneously meet their technology needs, benefit the environment, and realize dollar savings over the products' lives. Using the success of the Energy Star program as a precedent, the federal government has taken steps to encourage environmentally preferable choices. We also applaud federal agency and facility donation and recycling practices for providing valuable learning tools to thousands of school children while, at the same time, providing at least some protection against their equipment ending up in landfills or overseas. Such programs have also demonstrated that relatively simple and inexpensive steps can help ensure that donated and recycled e-waste is ultimately managed in a responsible manner. In particular, the FEC provides a framework through which participants can help ensure responsible recycling through downstream auditing of recipient organizations' disposal practices and by following guidance on how to select responsible recyclers. The federal government has the opportunity to lead by example and to leverage its substantial market power by broadening and deepening agency and facility participation in EPA electronic product stewardship initiatives, but meaningful results will only occur if federal agencies and facilities fully participate and utilize these promising initiatives' resources.

Ms. Chairwoman, this concludes my prepared statement. I would be happy to respond to any questions that you or other Members of the Subcommittee may have at this time.

End Notes

[1] GAO, *Electronic Waste: Strengthening the Role of the Federal Government in Encouraging Recycling and Reuse*, GAO-06-47 (Washington, D.C.: Nov. 10, 2005).

[2] GAO, *Electronic Waste: EPA Needs to Better Control Harmful U.S. Exports through Stronger Enforcement and More Comprehensive Regulation*, GAO-08-1044 (Washington, D.C.: Aug. 28, 2008).

[3] For updated EPA data, we examined EPA's procedures for accurately entering federal agency- and facility-provided data into its database, synthesizing the data, and using them for any calculations; we also interviewed EPA staff on steps they take to ensure the reliability of the data. We believe the data reported to EPA are sufficiently reliable for the purpose of updating information from our prior work.

[4] "Why is Total Cost of Ownership Important?" John Taylor Baily and Stephen R. Heidt. Darwin Magazine, November 2003.

[5] GAO-08-1044.

[6] GAO-06-47.

[7] FEC is sponsored by EPA and the White House Office of the Federal Environmental Executive.

[8] Resources include instruction sheets, tips, and checklists, among other things, which participants can choose to use.

[9] If a federal agency or facility chooses to achieve gold-level participation in FEC, it must document that for all electronics recycling it used EPA-preferred recyclers, such as the recycling electronics and asset disposition services, federal prison industries (UNICOR), a manufacturer's take-back service for EPEAT-registered electronics, or an electronics recycler that the participating agency or facility has conducted a physical on-site review of in the last 3 years.

[10] Some facilities within the Department of State, the Social Security Administration, and the National Aeronautical and Space Administration participate in the FEC, but these agencies have not registered with the FEC signifying participation.

[11] EPA generated these results using agency- and facility- reported data entered into the agency's environmental benefits calculator, which was developed to assist organizations in estimating the environmental and economic benefits of "greening" their purchase, use, and disposal of electronics. EPA posted these results on its FEC Web site.

[12] Executive Order 13514, "Federal Leadership in Environmental, Energy, and Economic performance," Oct. 5, 2009, reiterates the requirement that agencies purchase EPEATregistered electronics.

[13] Two facility participants from two agencies received an FEC gold award; 10 facility participants from five agencies received an FEC silver award, and 18 facility participants from seven agencies received an FEC bronze award.

[14] GAO-08-1044.

[15] EPA's environmental benefits calculator was developed to assist organizations in estimating the environmental and economic benefits of "greening" their purchase, use, and disposal of electronics.

[16] In addition, procurement using EPEAT criteria would lead to substantial reductions in emissions to air and water, as well as to the solid waste stream.

In: Environmental Considerations in Federal Procurement
Editor: John G. Kaspin

ISBN: 978-1-61209-532-5
© 2011 Nova Science Publishers, Inc.

Chapter 4

STATEMENT OF CASEY COLEMAN, CHIEF INFORMATION OFFICER, U.S. GENERAL SERVICES ADMINISTRATION, BEFORE THE SUBCOMMITTEE ON MANAGEMENT, ORGANIZATION AND PROCUREMENT

Good morning Chairwoman Watson, Ranking Member Bilbray, and members of the Subcommittee. My name is Casey Coleman and I am the Chief Information Officer of the U.S. General Services Administration. Thank you for inviting me to appear before you today to discuss the application of green policies in the life cycle management of GSA's Information Technology (IT) assets.

GSA has taken a lifecycle approach to our sustainability program. We buy energy efficient devices certified through the Electronic Product Environmental Assessment Tool (EPEAT), a program developed with a grant from the Environmental Protection Agency. We encourage our users to rely on fewer devices, such as shared multifunction printers rather than a printer on every desk. We employ modern approaches such as virtualization to reduce the number of servers required to perform the same work. And we have a program to ensure that assets at their end of life are recycled responsibly.

GSA's path toward a sustainable, Green IT program began maturing in 2007. That year, GSA began a program to consolidate all agency infrastructure and operations into one program, called GSA IT Global Operations (GITGO). We consolidated 39 contracts and 15 helpdesks into a single program, under the management of the GSA CIO. The GITGO program was critical to enabling our Green IT efforts. Previously, our IT assets were not standardized and there was no central accountability or visibility into where we had assets deployed. Through GITGO, the Office of the Chief Information Officer (OCIO) began to modernize, consolidate, and standardize the agency's infrastructure. Our goals were cost savings, improved sustainability, and equipping our workforce with modern tools, along with effective support procedures, necessary for them to perform their mission effectively.

Our infrastructure management efforts have yielded significant green benefits in several areas.

Server consolidation: One important GSA-wide initiative was the modernization and consolidation of servers. In this initiative we modernized all of the agency's Local Area Network (LAN) and Citrix servers. Some of these servers were six to eight years old. By purchasing new, more powerful servers, virtualizing them, and centralizing operations (rather than the old model of operating every function in every region and every organization), we have shut down over 700 servers yielding green benefits in areas from lower electricity use to a smaller footprint and reduced overhead.

GSA Telework Challenge: In 2007 GSA undertook an ambitious program to increase the number of employees regularly teleworking. The benefits of telework are well known—increased employee morale and engagement, reduced roadway congestion, reduced dependence on foreign oil, and increased ability to operate in the event of a Continuity of Operations situation. Telework also has green benefits, due to employees working at home or a nearby telework center, thus forgoing a long daily commute which reduces greenhouse gases emitted from automobiles as well as reducing the amount of gasoline consumed in commuting. The baseline from which GSA started in 2007 was less than 20% of its employees teleworking. We set ambitious goals for percentages of employees teleworking: 20% in 2008, 40% in 2009, and 50% in 2010.

In order to support the Telework Challenge, my office began an agency-wide workstation refresh program. We started with the organizations with the oldest and least efficient machines, and replaced them predominately with laptops, which are 20% more energy efficient than the machines they replaced. We implemented a telework training program and rolled out aids to help our employees make a successful migration to telework. After two years of this refresh, we are ahead of our goal. At last count, 46% of eligible GSA employees were teleworking.

Printer consolidation: Upon reviewing the deployment and use of printers within GSA, my office discovered that we had a ratio of almost one printer to every two employees, far greater than industry best practices of one printer for every 10 to 12 employees. We also had hundreds of different models, creating a very difficult support environment. To address this situation, I implemented a policy that encourages employees to rely on fewer devices, such as shared highspeed printers instead of personal printers on the desktop. The average printer is used less than 15 minutes per day, yet is powered on 24/7. In the past year under the initiative, GSA's number of printers has dropped 7%, which indicates that we are on the right track. We are also moving towards digital document management, which will ultimately reduce the number of printouts and paper that we consume. Both of these efforts have very tangible and immediate green benefits by reducing both electricity use as well as the amount of waste generated.

GSA has also implemented a strong series of initiatives to ensure that our IT assets are disposed of in an environmentally friendly and green manner. First, this is done by ensuring that no IT assets are simply thrown away. Instead, IT asset are disposed of through programs that allow for continuing use of the devices where possible. IT assets are first offered to other Federal agencies, where the agency requests the items. Next, a large percentage of IT assets for which GSA no longer has a need are given to schools and other non-profits through programs such as Computers for Learning. Finally, IT assets are put up for public auction, allowing the general public to bid on, and reuse, these devices. All these processes result in fewer IT assets going into landfills and help meet community needs.

Another area where GSA is ensuring environmentally friendly disposal of IT assets is with toner cartridges. We ensure that used printer toner cartridges are recycled, rather than ending up going to landfills. GSA's contract for printer services requires the contractor to provide prepaid return services and to ensure that toner cartridges are recycled and remaining toner is recycled or disposed of in a manner that complies with all environmental and human health and safety laws.

In addition to the major initiatives I have discussed, GSA has implemented more mundane, but very important, procedures to green our IT operations. For example, we have implemented power settings on all of our laptop and desktop computers which switch those devices into standby or sleep mode when not in use. This simple change saves a large amount of electricity when multiplied across a large organization like GSA. We are also putting into place smart power strips that can be used to end electricity drain by so-called "vampire" devices, those devices that continue to use electricity even when turned off. On the broader scale, we are investigating new technologies, such as cloud computing and green data center advances, which offer the promise of further significant reductions in energy consumption.

Chairwoman Watson, Ranking Member Bilbray and members of the committee, thank you again for the opportunity to appear before you to discuss this important subject. GSA is committed to environmentally friendly policies and procedures throughout the IT lifecycle. We will continue our current initiatives and are constantly on the lookout for new and innovative ways to get even greener. This concludes my testimony, I will be happy to answer any questions you may have.

In: Environmental Considerations in Federal Procurement
Editor: John G. Kaspin

ISBN: 978-1-61209-532-5
© 2011 Nova Science Publishers, Inc.

Chapter 5

TESTIMONY OF JAMES JONES, DEPUTY ASSISTANT ADMINISTRATOR FOR PREVENTION, PESTICIDES AND TOXIC SUBSTANCES, BEFORE THE SUBCOMMITTEE ON MANAGEMENT, ORGANIZATION AND PROCUREMENT

Good morning, Madam Chairwoman and Members of the Subcommittee. I am James Jones, Deputy Assistant Administrator for Prevention, Pesticides and Toxic Substances at the United States Environmental Protection Agency. I welcome this opportunity to discuss EPA's role in the procurement and management of the Federal Government's green initiatives related to IT assets.

The Environmental Protection Agency has taken a broad and ambitious approach to greening IT procurement in the Federal Government's purchasing practices, in the private sector in the US, and -- I'm pleased to note -- internationally, as well. As the nation's environmental agency, EPA has a leadership role in a number of programs designed to reduce environmental impacts across the full life-cycle of product manufacture, purchasing, use and disposal.

These programs help us meet – and where possible, exceed – the mandates of *Federal Acquisition Regulations* pertaining to green procurement, as well as several Executive Orders focused on pollution prevention, including Executive Order 13514 on *Federal Leadership in Environmental, Energy, and Economic Performance*, issued by the White House earlier this month. This latest Executive Order sets ambitious environmental goals for federal agencies, with strong accountability and transparency measures, and includes a requirement to *"leverage federal purchasing power"* in order to promote green products.

EPA is involved in a number of key programs that have helped us on the path to making electronics procurement more sustainable: **Energy Star**, **EPEAT** -- the *Electronic Product Environmental Assessment Tool* -- **FEC** -- the *Federal Electronics Challenge*, and **READ** – the *Recycling Electronics and Asset Disposition* program. I'll describe the programs in just a moment, but first, I'd like to mention some of the broad principles that guide our actions here.

- The first is **sustainability**. We are looking to reduce the environmental footprint of information technology -- equipment like computers, laptops, and monitors -- over the entire life-cycle of these products. From the time materials are first extracted from the earth through mining or drilling, to the end of the useful life of a computer product, our aim is to foster stewardship and reduce overall environmental impacts. This multi-attribute focus includes:
  - minimizing greenhouse gas emissions,
  - less reliance on toxic materials such as lead and mercury,
  - increasing use of recycled materials in manufacturing and assembling the product, and increasing the recyclability of components once the product is disassembled,
  - increasing material and energy efficiency --reducing the need for material disposal
- Second, we are committed to **building the partnerships** needed to achieve genuine and lasting results. EPA works closely with electronic manufacturers, standard-setting organizations, environmental and community groups, trade associations, states, and of course, other federal agencies, to create a broad consensus around sustainability, and build the framework for rigorous achievements in a partnership setting. We have extended these partnerships to the international community, as well, and are beginning to see global-scale results of our collective efforts.
- Third, I want to mention the importance of **transparency and accountability**. We are committed to an open, well-documented process where anyone can not only view program results, but can "drill-down" to get additional details as desired, and can view the modeling parameters that underlie the calculations of program accomplishments.
- The last key principle is **effectiveness**. These programs have realized substantial environmental results -- results that are increasing rapidly over time -- with only a very modest commitment of taxpayer dollars.

Let me take a few minutes to briefly describe these programs.

ENERGY STAR

ENERGY STAR is a joint program of EPA and the U.S. Department of Energy helping save money and protect the environment through energy efficient products and practices. Since its inception in 1992, the ENERGY STAR program has overcome many market barriers and helped revolutionize the marketplace for cost-effective, energy-efficient products and services. The program is a trusted source of unbiased information that helps homeowners, businesses, and other consumers understand their opportunities for energy savings and identify the reliable, cost-effective, efficient products and services that capture these savings.

The ENERGY STAR program focuses on driving greater efficiency in the following areas:

- Helping consumers identify new energy-efficient products that operate well beyond federal minimum efficiency requirements across more than 60 product categories for the home and office.
- Constructing efficient new homes and commercial buildings—public housing, multifamily and single family housing, schools, office buildings, hospitals, hotels, and others—that exceed code and meet rigorous benchmarks for energy efficiency.
- Improving the efficiency of existing homes, commercial buildings, and industrial facilities through standardized measurement systems, proven energy management strategies, and new energy efficiency services that overcome lingering market barriers.

Through 2008, more than 15,000 organizations have partnered with ENERGY STAR. They have achieved and helped the country achieve significant environmental and financial benefits.

Results are already adding up. Our partners, with the help of ENERGY STAR, have reported saving enough energy in 2008 alone to avoid greenhouse gas emissions equivalent to those from millions of cars — all while saving billions on their utility bills.

EPEAT – which I'll talk about next – and ENERGY STAR are closely entwined. Products must meet ENERGY STAR requirements in order to be EPEAT registered. ENERGY STAR staff participate in the development of EPEAT criteria, and the EPEAT and the ENERGY STAR programs co-market their work to federal purchasers.

EPEAT

EPEAT, helps purchasers identify and buy environmentally preferable electronics, and helps manufacturers gain market advantage by building greener electronic products. EPA supported the development of EPEAT -- providing grants, staff expertise, administrative and financial support – to respond to the needs of the marketplace. Purchasers wanted a definition of an *environmentally preferable electronic product*, and a list of products meeting that definition. Manufacturers needed a way to get credit in the marketplace for going the extra mile to reduce the impact of the products they create.

EPEAT is comprised of three main components:

1. An IEEE voluntary environmental performance standard that defines "green" for computer desktops, laptops, and monitors,
2. a registry of products meeting the criteria laid out in this standard, and a rigorous verification process, both managed by the Green Electronics Council, and
3. a calculator that determines the environmental benefits of each purchase of an EPEAT registered product.

EPEAT provides a marketplace 'reward' -- recognition as an EPEAT Bronze, Silver or Gold product -- for computers, laptops and monitors that meet EPEAT's stringent standards.

You can think of EPEAT as building on the successes of the ENERGY STAR program. In addition to meeting ENERGY STAR requirements, EPEAT registered products adhere to a total of 23 mandatory criteria adopted by the *Institute of Electrical and Electronics Engineers* as a National Standard (*IEEE 1680*). Participants who want Silver or Gold recognition can achieve it through adherence to some of the 28 optional criteria that are also included in the standard.

The criteria in the standard target many different environmental end-points, from energy use, to reduction or elimination of toxic metals and chemicals, and even product packaging.

The EPEAT program has had a tremendous and enthusiastic response. There are almost 3,000 EPEAT-registered products from 32 manufacturers. Think of any well-known computer manufacturer -- Apple, Dell, HP, Toshiba -- and they are EPEAT participants. EPEAT registered computers are now at airport screening stations across the country, and in every Kaiser Permanente hospital room, just to name a few purchasers of EPEAT products.

In 2008, according to the Green Electronics Council, purchases of 44 million EPEAT products in the US realized the following benefits:

- Reduced use of toxic materials, including mercury, by 1,021 metric tons
- Avoided the disposal of 43 thousand metric tons of hazardous waste
- Eliminated 14,353 metric tons of solid waste
- Saved over 8.39 billion kilowatt-hours of electricity — enough to power over 700,000 US homes for a year
- Reduced more than 1.57 million metric tons of greenhouse gas emissions — equivalent to taking over one million US passenger cars off the road for a year

In fact, users will save an estimated $794 million over the life of these products, chiefly from reductions in energy demand, in large part due to EPEAT products being required to meet the ENERGY STAR specifications.

It's not just the US market that realizes the environmental and economic benefits from EPEAT-registered equipment. Recently, international participants include most of Europe, Japan, and huge growing markets like China and Brazil. EPEAT is seeing environmental benefits in international markets on a similar scale to those I just described for the US, and is taking steps to more fully quantify results outside the US.

Given the enthusiastic participation thus far, EPA is actively exploring opportunities with the many EPEAT partners to expand the program beyond its current slate of computer products. Standards are being created for copiers and fax machines, and televisions, and plans are in place to develop standards for servers and cell phones. The program is also working with retailers and manufacturers to develop a plan to more actively market EPEAT products to consumers interested in procuring environmentally preferable electronics.

FEC

The Federal Government, perhaps the largest purchaser of consumer products in the US, spends an estimated $74 billion a year on information technology, according to 2009 OMB estimates.

EPA, working with the *Federal Environmental Executive*, recognized this as an enormous opportunity for the Federal Government to green its own house, and set an example on ways to purchase greener electronic products, reduce their impacts during product use, and manage obsolete electronics in an environmentally safe way.

Accordingly, we invited federal facilities to participate in the *Federal Environmental Challenge*, to meet the following goals by the end of 2010:

- 95% of computer purchases are EPEAT-registered, and 100% are Energy Star enabled.
- Electronic equipment achieves an average life span of at least four years.
- 100% percent of non-reusable electronic equipment is recycled using environmentally sound management.

Sixteen federal agencies and more than 220 federal facilities are participating in the FEC, and are on track to meet most of the 2010 goals. For instance, FEC partners reported that 88% of computers and monitors in 2008 were EPEAT registered... getting close to the 95% goal. The average life of computer equipment in 2008 was 45 months, close to the 4-year (48-months) goal for 2010.

Other goals are more of a challenge. While most FEC computers and monitors are ENERGY STAR equipment, not all equipment has energy-savings features fully enabled.

Thus far, operating with a budget of only $50,000, EPA support to the FEC has helped achieve the following results in 2008:

- 426,181 megawatt-hours in energy savings
- Over 100,000 metric tons in primary material savings
- Over 31,000 metric tons of greenhouse gases avoided, and over a million metric tons of air pollution prevented
- Almost 4,000 tons of discharges to waterways prevented

And, I'm pleased to report, an estimated overall savings of about $40 million in 2008 alone, from reduced energy and resource use.

READ

The Federal Government manages the disposition of about 10,000 computers a week, not to mention other forms of electronics. In order to assist federal agencies in recycling as much of these materials as possible, and safely disposing of the remainder, EPA manages READ – the *Recycling Electronics and Asset Disposition* program.

Pursuant to the 1996 Clinger-Cohen Act, OMB granted EPA the authority, in 2004, to create a *Government Wide Acquisition Contract* -- a single contract that can be used by all agencies – to handle recycling and disposal of used electronics.

The READ program developed this contract in 2005, which provides Federal agencies with a dependable method of properly managing electronic inventories, recycling electronic equipment, and disposing of excess or obsolete electronic equipment in an environmentally responsible manner.

EPA has awarded contracts to companies that can evaluate each piece of unwanted equipment and its components, and then reuse, recycle, or dispose of them under the following hierarchy:

- Refurbish and resell them, using the proceeds to offset costs.
- Donate them to charitable causes.
- Recycle as much as possible.
- Properly dispose of the remainder.

In addition to providing environmentally responsible disposition of electronic assets, READ also provides data security and economic value. READ services include:

- Developing a reportable audit trail of the equipment's final destination;
- Addressing appropriate levels of security for sensitive electronic data;
- Maximizing potential revenues from usable electronic equipment through a share-in-savings program.

The READ program received the 2006 White House *Closing the Circle Award* for environmental excellence in the Federal Government.

These program descriptions and results just scratch the surface of what is taking place in the world of greening IT in the Federal Government, but I trust they provide a good sense of the opportunities before us.

I'll be glad to respond to any questions you may have.

In: Environmental Considerations in Federal Procurement
Editor: John G. Kaspin

ISBN: 978-1-61209-532-5
© 2011 Nova Science Publishers, Inc.

Chapter 6

Testimony of Rick Goss, Vice President for Environment and Sustainability Information Technology Industry Council, before the Subcommittee on Management, Organization and Procurement, regarding "IT Procurement and Disposal: Application of the Federal Government's Green Policies in the Life Cycle Management of its IT Assets"

Introduction

Thank you Chairwoman Watson, Representative Bilbray and Members of the Subcommittee for the opportunity to testify regarding the federal government's green policies related to the life-cycle management of its IT assets. My name is Rick Goss and I am the Vice President of Environment and Sustainability for the Information Technology Industry Council, or ITI. ITI is a trade association representing 43 major high-tech and electronics manufacturers in the information and communications technology (ICT) sector. Our members are global leaders in all facets of ICT innovation, from hardware to services to software. ITI promotes policies that favor innovation, sustainability and open market competition. Through our Environmental Leadership Council, we also represent several dozen additional prominent manufacturers in the high-tech arena.

Our member companies have long been leaders in sustainability: many exceed the requirements on environmental design and energy efficiency, and lead the way in product stewardship efforts. As a result, the Dow Jones Sustainability Index, the Financial Times Sustainability Index, and the Global 100 have consistently recognized numerous ITI members for their concrete environmental and sustainability achievements. In addition, *Newsweek* recently issued its 2009 Green Rankings of America's 500 largest corporations, based on their overall environmental performance, policies, and reputation. Our members occupy 4 of the top 5 positions, and 14 of the top 40 slots overall.

On behalf of our member companies, I am pleased to testify today regarding the ongoing environmental achievements of our companies and to offer our suggestions regarding federal government policies, practices and opportunities to improve IT life-cycle management. The federal government is the world's largest single purchaser of ICT equipment and services, and is expected to spend over $75 billion in 2010. Consequently, the federal government has an unparalleled role to play in terms of driving continuous sustainability achievements in the ICT marketplace and in setting a leading example throughout the public and private sectors.

I have divided my remarks and recommendations into two sections. The first focuses on the environmental attributes and life-cycle management considerations of ICT products procured by the government. The second addresses the broad energy, climate and economic benefits that the federal government can drive through its own procurement of advanced ICT systems and services, and by enacting policies that encourage ICT deployment throughout the broader economy.

In addition, I would be remiss in my testimony if I did not introduce the concepts of cloud computing and virtualization to the Subcommittee as a potential paradigm-shift in how the federal government procures ICT systems and services. I would recommend that the Subcommittee consider hosting a follow-up hearing to explore cloud computing and virtualization innovations and their impacts on future government enterprise ICT procurement.

SECTION 1. ICT PRODUCT ACHIEVEMENTS & CONSIDERATIONS

ICT manufacturers recognize that we have a critical role to play in the sustainability effort by continuously improving product environmental, energy and performance characteristics. Our companies have consistently risen to that challenge. As a result of our members' abiding dedication to product stewardship and technological innovation, the high-tech and electronics industries continue to achieve significant and sustained environmental innovation throughout the entire product lifecycle: from environmental design to energy efficiency, beneficial reuse and proper end-of-life management.

It is also critical to emphasize that the competitive marketplace – rather than government mandates and regulation – continues to be the primary driver behind these improvements. On the whole, every year our products become more energy efficient, incorporate innovative and environmentally-preferable materials, and become easier to upgrade, disassemble and recycle. This process of continuous evolution, driven by market demand and competition, can be readily observed by comparing today's products to similar products that were manufactured just a few years ago. The federal government can and should continue to exert a major market influence.

Energy Star & EPEAT

Two of the federal government's key programs in the areas of green IT procurement are Energy Star and EPEAT - The Electronic Product Environmental Assessment Tool. ITI and our members companies are strong supporters and active participants in both of these key initiatives. Both programs promise valuable and concrete market rewards to those leadership companies that make the significant time and resource investments necessary to ensure that our products meet or exceed the requirements.

For purposes of today's hearing, the Energy Star program includes such key products as computers, copiers and fax machines, monitors, printers and scanners, and servers. For almost two decades, Energy Star has been the premier product energy efficiency benchmark, and is well recognized by public, private and individual purchasers alike. U.S. EPA reports that, since 2000, Americans have purchased more than 2.5 billion Energy Star qualified products, resulting in significant cost and energy savings. Over 80% of products sold are in the categories of home and office electronics. Energy Star is recognized globally as the de-facto standard for energy performance, and its strength is in the simplicity of its purpose and recognition as a brand.

The EPEAT program is based on IEEE 1680, a multi-attribute international standard developed by a diverse group of stakeholders through a consensus-based process. The EPEAT program was developed to allow IT purchasers to base procurement decisions on the full life-cycle environmental attributes of high-tech products. EPEAT incorporates recognized metrics and criteria, which can be measured, verified and consistently applied. The EPEAT standard currently applies to computers and monitors, and is currently being expanded to include (among other products) imaging equipment – another major product category purchased in large volume by the federal government and other major institutional purchasers.

Currently, federal, state and local governments, foreign governments, commercial entities and academic institutions have specified EPEAT in over $60 billion worth of procurement contracts. Members of ITI and our Environmental Leadership Council dominate the EPEAT registry, and presently manufacture 90% of the 419 EPEAT Gold- rated products.

Energy Star and EPEAT Recommendations

ITI would like to offer the following recommendations in relation to Energy Star and EPEAT:

1. **Increase Federal Government Procurement of Energy Star and EPEAT Products.** President Obama's new Executive Order on "Federal Leadership in Environmental, Energy, and Economic Performance" includes specific provisions on procuring Energy Star and EPEAT electronic equipment. This reinforces an Executive Order issued by President Bush in 2007. ITI supports this commitment, and looks forward to working with the Administration and Congre ss on its fulfillment.
Purchasing requirements for Energy Star and EPEAT products are already included in the Federal Acquisition Regulation. However, previous studies have demonstrated that federal agencies are not purchasing Energy Star products as directed. This lack of compliance has resulted in a reported annual taxpayer energy expense of over $400 million, and associated emissions of approximately 3 million metric tons of carbon dioxide. Since Energy Star is a mandatory criterion in the EPEAT program, this means that purchases of EPEAT-registered products are also suppressed. This outcome also significantly undermines the market incentive for manufacturers to participate in these programs. ITI and our member companies stand ready to help

implement this Executive Order and improve federal purchases of environmentally preferable products.

2. **Provide Targeted Funding for Education and Outreach Efforts.** The largest and most immediate opportunity to secure additional energy and cost savings is by educating purchasers about the benefits of buying Energy Star and EPEAT qualified products. This includes informing purchasers regarding how to take full advantage of the energy-saving capabilities already designed into nearly every ICT product in use today. Congress should consider funding for initiatives to promote broader awareness on energy efficiency and the Energy Star Program, on effective use of power management features incorporated into ICT products and systems, and on the life-cycle benefits of purchasing EPEAT products.

3. **Maintain the Focus of the Energy Star Program on Product Energy Use.** U.S.EPA is considering a potential expansion of the Energy Star program to factor "additional energy impacts" into the specification. This proposal would result in a wide range of new and unclear factors and detract from the traditional focus of the Energy Star program on the energy consumption of products.

   The consideration of additional energy impacts would necessitate a complex evaluation of a multitude of factors in order to account for the full picture of related energy impacts associated with a product. Among others, these would include the energy impacts associated with:
   - Manufacturing operations
   - Supply chain management
   - Recycled or recovered content in products
   - Product distribution
   - Product packaging
   - Telework programs for employees
   - Product recycling programs

   The process of quantifying and measuring these impacts would be highly complex, and would prove confusing for businesses and consumers. The success of the Energy Star label is due in large part to the fact that it is objectively measurable and verifiable. By focusing solely on the attribute of energy consumption of a particular product model, Energy Star offers product purchasers a straightforward and objective means of evaluating the energy efficiency of a product. Including other environmental considerations will dilute this benefit and introduce the consideration of complex impacts based largely on poor data quality. Energy Star would evolve into a multi-attribute eco-label based on vague criteria and an unknown means of verification.

   EPA should maintain the focus of the Energy Star program solely on the energy consumption of products as used by the purchaser, without regard for other "additional energy impacts."

4. **Provide Additional Funding and Oversight of EPEAT.** Manufacturers want to see EPEAT succeed and become the premier international procurement program for green electronics. ITI and our member companies have devoted hundreds of hours to developing this standard with other stakeholders, redesigning products to meet

criteria and working with U.S. EPA and EPEAT staff to strengthen and improve the program.

EPEAT was created as an institutional procurement tool, with the federal government as its primary audience. U.S. EPA is a major EPEAT stakeholder and, since the program's launch in 2007, has actively encouraged its expansion into additional countries, into the consumer marketplace and into several additional product categories. The program is expanding at a rapid pace, but is suffering from a lack of adequate funding and oversight.

The administration of the EPEAT program is currently funded solely by manufacturer registration fees. In addition, manufacturers are providing financial support for the development of the new EPEAT standard for imaging equipment. U.S. EPA funded the development of the original EPEAT standard for computers and monitors and provided a small grant to launch the program. While the agency has provided partial support for the development of four new standards, there has been no additional support to revise the original standard or provide for the long term stability of this important program.

In addition, ITI and our members strongly support additional federal government oversight of the EPEAT program. EPEAT is a federal procurement requirement and is being increasingly recognized in international venues and by numerous private purchasers. However, the EPEAT program is managed by a small, third- party, single source provider. ITI encourages Congress to provide additional funding and oversight of EPEAT to make sure this important program succeeds.

Additional Product Recommendations

1. **Ensure Harmonization of Procurement Requirements Between Federal Agencies.** As federal agencies seek to implement the new Executive Order, the federal government should ensure that procurement requirements are consistent. A lack of consistency can lead to a fragmented federal marketplace, with different contract specifications and paperwork requirements. This could result in delays and increased costs that neutralize the advantages for manufacturers that meet or exceed the requirements.
2. **Rely on Advanced Printing Solutions.** New digital and two sided thermal printing technologies increase printing speeds and avoid print overruns while saving energy, paper and consumables. By relying on networked printers and digital printing solutions, public and private sector entities alike can achieve significant cost savings while increasing efficiency and reducing associated carbon emissions. According to industry estimates, for every one dollar of actual printing, there are $5 -8 of other costs, including warehousing, distribution and inventory obsolescence. Each of these has an associated energy use and emissions footprint, and each can be reduced by switching to "on-demand" digital solutions. The carbon abatement potential due to minimizing overruns of books, magazines and newspapers through digital printing services is on the order of 114 to 251 million tons of CO_2 equivalent (this estimation includes a rough estimate of the embedded carbon and energy use of the printers as well as the paper production).

3. **Increase the Use of Video Conferencing.** ITI members have pioneered the development of video conferencing technologies that make virtual meetings possible. Video conferencing improves productivity while avoiding thousands of miles of car and air travel and associated carbon emissions. The federal government can achieve significant cost and energy savings and play an important role in the deployment and broader adoption of these technologies by increasing its own use of video conferencing solutions.

Additional Product Considerations

The Subcommittee has indicated that it has a particular interest in learning more about the use of recycled materials in ICT products and about proper management of used and obsolete devices.

Recycled Content

ITI member companies use significant quantities of recycled materials, including glass, metals and plastics, in new generations of our products, thus creating demand that helps sustain markets for these materials. To qualify for EPEAT, at least 65 percent of the materials in the product must be reusable or recyclable using current infrastructure and technologies. In addition, our companies have continuously decreased the volume of packaging materials while simultaneously increasing the percentage of recycled content.

Asset Management & Recycling

ITI members offer comprehensive asset management and product recycling services to the federal government as well as to our commercial and institutional customers. In fact, our companies have provided for the proper recovery and management of well over two billion pounds of used electronics products. Our members or their authorized contractors recover used devices, transfer data to new systems, remove sensitive information and properly manage recovered equipment. Functional equipment is typically refurbished and returned to commerce for reuse. This is an environmentally-beneficial practice that helps derive the most value from the resources used to manufacture the equipment. Our members ensure that older or broken units are first used for spare parts, as appropriate, and then recycled in an environmentally-sound manner, with any commodities reclaimed for subsequent reuse.

ITI and our member companies have been working in close coordination with U.S. EPA and other stakeholders to develop a set of Responsible Recycling practices for electronics recycling. The R2 practices are intended to ensure that obsolete electronics are managed and recycled in a safe and environmentally-appropriate manner that is protective of human health and the environment. R2 is in the process of becoming a recognized standard, which will allow government, commercial, institutional and residential consumers alike to know that their obsolete products will be properly managed.

On the consumer side, many of our leading members offer voluntary national programs at little or no cost to our residential customers. Unlike the government and commercial marketplace, where our companies often sell thousands of units under a single contract, the consumer marketplace is very fragmented, as sales are often made one at a time at retail. In addition, manufacturers must rely on a national infrastructure of wholesalers, distributors and retailers to transport our products (often across state lines) and deliver them for final retail sale. With few exceptions, we do not have a direct connection with our customers at the point of sale.

As part of our corporate commitments on environment and sustainability, manufacturers believe we have a role – though not the only role – to play in offering recovery and recycling solutions to consumers. Our members provide a variety of recycling options to consumers, which include a combination of collection centers, special collection events, and mail-back programs.

SECTION 2. THE ENERGY, CLIMATE AND ECONOMIC BENEFITS OF ICT SYSTEMS AND SERVICES

The ICT sector plays a critical role in helping address major strategic challenges, including energy security, economic competitiveness and the transition to a clean energy economy. ICT hardware and software innovations yield transformative energy efficiency improvements, enable revolutionary Smart Grid, Smart Building and Smart Transportation systems and drive state-of-the-art renewable energy technologies. The ICT industry has powered major gains in U.S. and global economic productivity, has created tens of thousands of high-paying jobs here at home, and connects communities the world over to vital economic, educational and health resources.

Consider the following examples:

- The American Council for an Energy-Efficient Economy (ACEEE) concluded that "[f]or every extra kilowatt-hour of electricity that has been demanded by ICT, the U.S. economy increased its overall energy savings by a factor of about 10. These productivity gains have resulted in significant net savings in both energy and economic costs. The extraordinary implication of this finding is that ICT provide a net savings of energy across our economy."
- In a July 2009 report, ACEEE also concluded that energy efficiency measures are capable of delivering up to 50% of the emissions reductions necessary to meet the President's 2050 emissions reductions goals.
- The Smart 2020 report issued by The Climate Group concluded that ICT strategies for energy efficiency could reduce up to 15% of global emissions by 2020. The U.S. addendum to that report indicates that ICT strategies could reduce annual U.S. carbon emissions by up to 22% by 2020, which translates into energy and fuel savings of $140 to 240 billion dollars.
- The Information Technology & Innovation Foundation estimates that spurring an additional investment of $30 billion in America's IT network infrastructure would create approximately 949,000 U.S. jobs through expanded broadband networks, health IT and the Smart Grid.

Policy Recommendations

I have provided an addendum to my testimony that includes a number of suggestions to drive the accelerated deployment of ICT systems and services and help realize these clear benefits. I would also like to provide specific information in my testimony regarding two recommended policy approaches.

1. **Drive Energy Efficiency Considerations into Enterprise-Level Federal ICT Procurement.** EPA's *2007 Report to Congress on Server and Data Center Energy Efficiency* concluded that public and private sector data center managers rarely factor energy efficiency into their procurement and operational decisions:
    - "The barriers that prevent data centers from adopting changes that offer very reasonable paybacks are typically not technological but organizational. ... [one of these is] *Split incentives*: In many data centers, those responsible for purchasing and operating the IT equipment are not the same people that are responsible for the power and cooling infrastructure, who in turn typically pay the utility bills. This leads to a split incentive, in which those who are most able to control the energy use of the IT equipment (and therefore the data center) have little incentive to do so" (Page 11).
    - "Under this arrangement, most IT managers never see the energy bill for their equipment, and their job performance is not evaluated based on energy costs. While improved energy efficiency may benefit the organization overall, the data center manager will see little reward" (Page 86).
    - "In many organizations, it is standard practice to base IT equipment and software purchasing decisions on TCO [total cost of ownership], which includes the lifetime maintenance and support costs. These TCO calculations for IT equipment rarely include energy as a factor, but they should include both the cost of electricity to run and cool the hardware as well as the cooling, power conditioning, and other capital equipment required to support the IT hardware in the data center" (Page 109).

    Targeted federal action can help resolve this split-incentives issue, especially when it comes to innovations in areas such as advanced metering infrastructure, efficient data center strategies, applications modernization and rationalization, building systems energy efficiency, and telework. Therefore, ITI has been actively supportive of amendments by Representative Anna Eshoo and Senator Mark Udall that direct each federal agency to collaborate with OMB to create an implementation strategy for the purchase and use of energy efficient information and communications technologies and practices. Under this approach, each agency would (1) evaluate how ICT infrastructure could yield cost and energy savings; (2) meet new performance goals for energy efficient information and communications systems; and (3) be eligible to realize the savings and rewards brought about. The Eshoo Amendment is included in HR 2454 – the American Clean Energy and Security Act of 2009, and we strongly urge support for this provision. We are also hopeful that the President's new Executive Order will be implemented in a manner that helps address this problem, and look forward to working with the Administration to achieve this result.

2. **Direct the Department of Energy to Create Metrics to Determine the Annual Energy Rating of Enterprise-Level ICT Systems.** Energy efficiency is the most immediate and available method for reducing U.S. energy demand and increasing energy security. From "Smart" systems to advanced renewable energy technologies to electronic health records, policies in Washington will continue to rely more on ICT-led solutions to public policy challenges, thereby driving greater and more widespread adoption of ICT systems.

   Facing this trend, major ICT companies have unleashed a wave of hardware and software innovations to make each generation of systems more energy efficient than the last. Several public utility companies have quantified the dramatic energy savings associated with cutting edge technology that greatly improves the energy efficiency of enterprise ICT systems including Data centers, servers, network, and storage systems. With models that quantify energy savings associated with efficiency improvements in systems and configurations, these utilities can offer rebates to their customers that undertake ICT consolidation and virtualization projects. These rebates are based on the energy savings that will be achieved by the customer over a given time period. The efficiency improvements incentivized can reduce energy consumption in data centers by 60% or more.

   The primary barrier to the more widespread adoption of these programs is the absence of a recognized, government-approved methodology for quantifying these energy savings. Testing programs for ICT systems can be expensive to replicate, and many utilities lack the internal resources necessary to fully vet and substantiate industry best practices and create rebate programs. Moreover, industry participants are more likely to base economic incentives on methodologies that are validated by the DOE, rather than models developed by a private contractor or fellow utility. DOE should form partnerships with the utility companies to validate their ICT energy efficiency programs and to assist the utilities in sharing information and best practices. This will drive more meaningful demand mitigation programs that will yield these dramatic energy efficiency savings more quickly.

Thank you again for the opportunity to testify today on this important set of issues. I would be happy to respond to any questions.

Chapter 7

WRITTEN TESTIMONY OF JEFF OMELCHUCK, EPEAT EXECUTIVE DIRECTOR, BEFORE THE SUBCOMMITTEE ON MANAGEMENT, ORGANIZATION, AND PROCUREMENT

CONGRESSIONAL ACTION REQUESTED

- EPEAT is a sound green purchasing system for electronics that covers the complete lifecycle of environmental impacts, from toxics to packaging, including design for recycling. EPEAT is applying market-based forces to globally drive the design, manufacturing, and service practices of electronics makers to reduce impacts throughout the lifecycle, including making products more easily and efficiently recycled.
- The Federal Acquisition Regulation (FAR) and two Executive Orders require all agencies of the US Federal government to satisfy 95% of their need for electronics with products that are EPEAT registered. Yet many government contracts present catalogs of products to federal purchasers that include many non-EPEAT registered products and they do not identify which products are EPEAT registered, making it very difficult for purchasers to comply with federal regulation. **We urge Congress to require that government contracts and contractors clearly identify EPEAT registered electronics so that federal purchasers can more easily comply with federal purchasing regulations and Executive Orders.**
- The US government has for many years done a good job of using their own purchasing power to create demand for greener products. But this doesn't go far enough. Promoting green purchasing to the public based on sound lifecycle based programs educates consumers on what they can do and strengthens demand for environmentally preferable products. EPEAT is the program vetted by the US Government and used for its own purchasing of electronics. **EPA should support and promote EPEAT and other sound green purchasing systems to the public.**
- Developing an effective e-waste recycling program in the US is a critical environmental issue. However, recycling is fundamentally an "end of the pipe" activity that recovers perhaps a few percent of the environmental investment in

electronic products. To fully address the lifecycle impact of electronics, we must address them earlier in the product lifecycle. By specifying EPEAT, purchasers apply market pressure to change the design, manufacturing, and service practices of electronics makers globally and reduce impacts throughout the lifecycle, including making products more easily and efficiently recycled. **Buying greener electronics as defined by EPEAT should be recognized and supported as a necessary part of any long-term solution to e-waste.**

EPEAT OVERVIEW

- EPEAT is a "green purchasing system" for electronics. It is based on an open consensus-based standard that covers a full spectrum of green attributes, including energy efficiency, reduced toxics, design for recyclability, product longevity and sustainable packaging, and requires manufacturers to have take- back programs for product, batteries, and (optionally) packaging. Products are rated Bronze, Silver or Gold - green, greener, greenest.
- "The EPEAT System" is a growing complex of people and organizations working collaboratively. The "green standards" used in EPEAT are developed by hundreds of independent experts working in a formal public standards development process administered by The Institute for Electrical and Electronic Engineers (IEEE), an ANSI accredited standards body. EPEAT Inc. acts as the hub of the system and manages the web based product registry, product verification, and marketing the system to purchasers.
- EPEAT Inc. is an independent non-profit organization. It is not a program of US EPA or any other government agency. EPEAT now has 3 full time staff, half a dozen part time contractors, and dozens of volunteers who donate their time to promote EPEAT, provide advice, etc.
- EPEAT has received significant support from US EPA throughout its development and start-up but EPEAT Inc. is now supported by fees paid by manufacturers to register their products in the system. EPA is currently providing some support for the development of IEEE "EPEAT green standards" for new products types but is not providing any funding support for operations, consumer outreach, or periodic updating of the product standards.
- ENERGY STAR participated in developing EPEAT, and EPEAT's primary energy efficiency criterion is compliance with ENERGY STAR. ENERGY. STAR is also providing significant staff support for the development of EPEAT energy efficiency specs for printer-type products and TVs.
- EPEAT is used by the US federal government and public agencies in many other countries, states, and cities and private companies globally to specify "green" when they write purchase contracts for IT.
- All agencies of the USG are required by Executive Order and the Federal Acquisition Regulations (FAR) to satisfy 95% of their requirement for electronic products with EPEAT registered products, where an EPEAT standard exists for the product category. Currently that covers desktops, laptops, workstations, thin clients and

displays. OMB reports that most agencies' compliance is good and improving but not perfect.
- The USG's purchasing power, combined with that of the other global users of EPEAT, has created a $60 billion market incentive for manufacturers to design and manufacture greener electronics. In combination with other global regulatory and voluntary initiatives, this aggregated purchasing power helps drive environmental innovation in the design, manufacturing, and service practices of electronics companies globally, from the largest multinational brands to small local brands.
- Green purchasing must be an important component of any solution to the e-waste issue. While recycling electronics responsibly is critical, recycling alone, even where manufacturers are required to participate in or underwrite end of life management, does not effectively change the design of electronics over time to reduce their environmental impact. A green purchasing system creates a market incentive for manufacturers to design and manufacture greener products that can more easily be recycled. **EPEAT is a sound green purchasing system for electronics and should be recognized and supported as part of the long-term solution to e-waste.**
- Public awareness of the environmental issues associated with electronics is growing. However, consumers are not yet generally aware of their role in reducing those impacts. In particular, consumers are not aware of how to identify greener electronics and how purchasing green electronics benefits them, their communities, and their planet.
- EPEAT does not have the resources for broad public promotion and the USG has provided no support for that. ENERGY STAR has a long and successful history of promoting the importance of energy efficiency in the consumer market, due to decades of significant and consistent government funding. EPEAT and ENERGY STAR continue to explore possible collaboration on public promotion.
- It is good public policy, and a very efficient use of public resources, for EPA to support public promotion of green purchasing based on sound programs like EPEAT, the program the government uses for their own purchasing. **EPA should provide EPEAT support for public promotion, development of new standards and updating of existing standards.**
- **EPEAT is a new and innovative model of a powerful way to change materials management and achieve sustainability goals that should be studied and copied in other industries.**

HISTORY AND DEVELOPMENT OF EPEAT

The Green Electronics Council is a 501(c)(3) nonprofit organization that manages EPEAT, the green purchasing system for electronic products. EPEAT was developed beginning in 2003 by a group of diverse volunteer stakeholders representing all stakeholder constituencies interested in electronics and the environment, including industry/ manufacturers, environmental advocates, private and public purchasers of electronics, researchers, recyclers, government staffers and others. The process was facilitated by the non profit Zero Waste Alliance, supported by a grant from US EPA (more info at http://www.zerowaste.org/epeat/index.htm). As the process of criteria development moved forward, the stakeholders decided to formalize the system and criteria they were developing as a public technical standard working through the Institute of Electrical and Electronic Engineers, and in April 2006 they released public standard IEEE 1680. The standard contains both the environmental performance criteria for personal computer products and the design of the EPEAT system itself, through which those criteria are applied to products. The "EPEAT Standard" contains 51 criteria – 23 required and 28 optional covering the product's entire life cycle, from toxics to energy efficiency to design for recyclability to packaging, company performance, and product and battery takeback. Products that meet the 23 required baseline criteria are rated EPEAT Bronze. Products that meet 50% of the additional criteria are recognized as EPEAT Silver and the greenest products, which meet 75% or more of the optional criteria, earn a rating of EPEAT Gold.

The Green Electronics Council was selected by EPEAT's stakeholder "Implementation Team" to manage the EPEAT system. With support from an EPA startup grant, in July of 2006 GEC launched the EPEAT registry at www.epeat.net, an online searchable database of products that are registered by their manufacturers as meeting each of the 51 criteria. On EPEAT's launch there were 3 participating manufacturers and 60 products. Today the EPEAT registry has separate product registries for each of 40 countries, with over 40 participating manufacturers, including all the leading global brands and a large number of small brands, and over 1200 products registered in the US alone. EPEAT is now entirely self supporting, funded by the annual fees that manufacturers pay to register their products.

In December of 2006 President Bush issued Executive Order 13423 requiring all federal agencies to satisfy 95% of their requirements for electronic products with products that are EPEAT registered. A year later the EPEAT purchase requirement was codified into the Federal Acquisition Regulations (FAR SubPart 23.7). In September of 2009 President Obama renewed the USG's commitment to buy EPEAT registered green electronics in Executive Order 13514. EPEAT registration is now required by purchase contracts from government agencies in Canada, New Zealand, Australia, Mexico, Brazil, Singapore, Thailand, Poland, and Lithuania, and on a large number of contracts from state and municipal agencies, universities and colleges, healthcare facilities and systems and private corporations globally.

In 2009, partially supported by an EPA grant, stakeholders began developing IEEE/EPEAT Standards for Imaging Equipment (printers, fax machines, copiers, etc) and for televisions. They are expected to complete that work in 2010 and those product types should appear on the EPEAT registry in late 2010 or 2011. Standards development processes for servers and then cell phones will follow.

EPEAT is a remarkable example of a public-private partnership that is greening the design of electronics products and related service offerings using market forces rather than regulation.

US GOVERNMENT USE OF EPEAT AS A GREEN PURCHASING STANDARD FOR IT

- **January 2007** Executive Order 13423 was signed by President Bush –This E.O. consolidates and strengthens five executive orders and two memorandums of

understanding related to environmental, energy, and transportation performance and accountability, and required all Federal Agencies to purchase 95% or higher EPEAT registered products in all eligible product categories.
- **December 2007** The FAR Council integrated a requirement for use of EPEAT into the Federal Acquisition Regulations (FAR) as an interim rule.
- **February 2009** The FAR interim rule became final.

USG COMPLIANCE WITH EXECUTIVE ORDER AND FAR RULING

OMB tracks each agency's compliance with FEC requirements, including the requirement to buy EPEAT registered products. OMB reported to the Office of the Federal Environmental Executive the following.

- **FY 2007 Federal Electronics Challenge Partner EPEAT Purchasing Results**
  - 80% of desktops, laptops and monitors purchased by FEC partners were EPEAT registered
  - Suppliers reported sales of over 1 million EPEAT registered products to the Federal Government
- **FY 2008 FEC Partner EPEAT Purchasing Results**
  - 88% of computer desktops, laptops and monitors purchased or leased by Federal Electronics Challenge participants were EPEAT registered. Of those products, 2% were EPEAT Bronze; 46% were EPEAT Silver, and 40% were EPEAT Gold
- **Federal Purchase Totals EPA estimates total Federal EPEAT purchasing to date (Fall 2009) at ~ 2 million units.**
  Results compiled from the 2008 OMB Scorecard and Federal Electronics Challenge reporting provide insight into Federal agencies' successful implementation of the FAR requirement
  - Thirteen of 22 Federal Agencies reported meeting or exceeding the goal of 95% or higher EPEAT purchasing in their 2008 IT acquisitions. Individual descriptions of these agencies' successes are listed below.
  - The remainder of the Agencies reporting showed significant success –with the average compliance rate between 65-75%

Agencies Meeting FAR EPEAT Requirement in full in FY 2008

Department of Veterans Affairs: 100% of the 290,623 Dell desktops and monitors leased by VA between September 2007 and December 2008 were EPEAT gold or silver products (i.e., 135,598 EPEAT gold desktops; 4,345 EPEAT silver desktops; and 150,680 EPEAT silver monitors).

Department of the Treasury: Of the 64,686 computer desktops, laptops/notebooks, and monitors Treasury purchased in FY 2008, a total of 64,491 (or 99.7%) were EPEAT registered products. The EPEAT compliant products were 68.5% Silver and 31.2% Gold.

Department of Energy: DOE purchased more than 50,000 EPEAT registered computers and monitors in FY08, constituting more than 96% of DOE's approximately 52,000 total purchases. All but 4% were Gold (49%), Silver (45%) or Bronze (2%).

Social Security Administration: 100% of the 24,673 desktop computers purchased by SSA were EPEAT registered Gold. All of the 26,948 LCD monitors purchased were EPEAT registered, 15,600 were EPEAT registered Silver, and 11,348 were EPEAT registered Gold.

Department of the Interior: DOI purchased through their Agency wide mandatory use IT Hardware Contract a total of 34,737 desktops, 17,548 laptops, and 4,237 monitors, i.e., a total of 56,522 EPEAT compliant units in FY 2006 2008.

Department of State - All of State's Global Information Technology Modernization Program (GITM) purchases in FY 2008 were EPEAT Silver or Gold products: 7,515 EPEAT registered Gold desktops and 5,370 EPEAT registered Silver monitors.

Department of Commerce: DOC purchased 6,618 computers, monitors, and laptops in FY 2008; 6,423 (or 97%) of these were EPEAT registered.

NASA: In 2008, the Outsourcing Desktop Initiative for NASA (ODIN) program purchased 12,256 EPEAT registered computers and monitors for NASA personnel.

Tennessee Valley Authority: TVA purchased 2,496 desktop computers, 3,500 monitors, and 1,017 laptops that were EPEAT registered, for a total of 99.3% EPEAT compliant purchases in FY 2008.

Office of Personnel Management: OPM added EPEAT requirements into the IT Procurement Authorization tracking system (ITPA). All of the CIO approved purchases were either Gold or Silver rated. A total of 1,541 desktop/laptops and 1,619 LCDs were procured in FY 2008.

General Services Administration: GSA purchased more than 600 EPEAT certified computers in FY08. All computers were 100% compliant with EPEAT purchases.

US Environmental Protection Agency: In 2008, EPA used an IT Blanket Purchase Agreement (BPA) that specifies EPEAT registered equipment for electronic purchases. EPA estimated that 99% of eligible electronics purchases were EPEAT registered. In FY09, EPA is standardizing its user provisioned computer equipment for headquarters employees and providing more than 12,000 EPEAT registered computers to its staff.

Department of Labor: DOL instituted a new requirement that any non EPEAT purchase be registered as an exception to purchasing policy. No exemptions were reported in FY 2008, indicating 100% compliance with the EPEAT purchasing requirement.

COMMERCIAL AND ENVIRONMENTAL IMPACTS OF EPEAT

Manufacturers that register their products in EPEAT are required to annually report to GEC the number of EPEAT registered products that they sell. In 2008 EPEAT participating manufacturers sold more than 44 million EPEAT registered products in the US alone. In future years GEC will track and report sales in all 40 supported countries.

By feeding the manufacturer provided data into an Electronics Environmental Benefits Calculator (EEBC) that was developed by University of Tennessee Center for Clean Products working on an EPA grant, GEC is able estimate the lifecycle environmental benefits that result from the purchase of EPEAT registered green products as compared to the purchase of conventional products.

2008 US purchases of EPEAT registered laptops, desktops, and monitors over conventional products will:

- Reduce use of toxic materials, including mercury, by 1021 metric tons, equivalent to the weight of 510,949 bricks
- Eliminate use of enough mercury to fill 149,685 household fever thermometers
- Preclude the disposal of 43 thousand metric tons of hazardous waste, equivalent to the weight of almost 22 million bricks.
- Eliminate 14,353 Metric Tons of solid waste, equivalent to the amount 7202 U.S. households generate in a year

In addition, due to EPEAT's requirement that registered products meet ENERGY STAR's energy efficiency specifications, these products will consume less energy throughout their useful life, resulting in:

- Savings of over 8.39 billion kWh of electricity — enough to power over 700,000 US homes for a year
- Reduction in use of 14.8 million metric tons of primary materials, equivalent to the weight of more than 114 million refrigerators
- Avoidance of 34.2 million metric tons of air emissions (including greenhouse gas emissions) and over 71,000 metric tons of water pollutant emissions
- Reduction of over 1.57 million metric tons of greenhouse gas emissions — equivalent to taking over one million US passenger cars off the road for a year

In addition to these benefits, reported global sales demonstrate EPEAT's potential for reducing the environmental costs of computing worldwide. Despite only 27% of participating manufacturers reporting their Canadian EPEAT sales and only 20% reporting their Rest of World sales, the estimated benefit of EPEAT sales to these regions is still significant:

- Reduction of 2.8 million metric tons of primary materials
- Elimination of over a million kilograms of toxic materials, including enough mercury to fill 157,311 household fever thermometers
- 16,297 Metric Tons of solid waste eliminated
- Greenhouse gas emissions equivalent to removing 2.3 million US cars from the road for a year

Click here to read the FULL EPAT 2008 Environmental Benefits Report

EPA FINANCIAL SUPPORT OF EPEAT

EPA has provided, and continues to provide, both financial and in kind support to elements of the "EPEAT System", as identified below:

- 2002 – 2006 EPA provided a cooperative agreement in the amount of $290,000 to the Zero Waste Alliance to support facilitation of the multistakeholder consensus process that developed EPEAT and resulted in publication of the IEEE 1680 standard and selection of an organization to manage EPEAT.
- 2006 – 2008 EPA provided a cooperative agreement in the amount of $420,000 to Green Electronics Council to launch EPEAT as a working commercial system. EPEAT is now financially selfsustaining.
- 2007 EPA provided inkind support for development of marketing materials.
- 2008– 2011 EPA provided a grant in the amount of $419,000 to University of Tennessee Center for Clean Products to support management/facilitation of the IEEE standards process to develop EPEAT green standards for Imaging Equipment (printers, copiers, fax, etc.), televisions, server computers, and cell phones. This work is now in process.

In addition, 2 halftime EPA employees continue to work significantly with EPEAT, and several EPA, DOE, and other federal employees sit on EPEAT advisory boards or participate in the standards development process to provide a Federal Government perspective.

EPEAT IN THE CONSUMER MARKET

EPEAT was developed primarily by and for purchasers who buy electronics on purchase contracts, and it has seen astonishingly rapid growth in that market. Manufacturers, retailers, and others now see EPEAT as a credible way to communicate the "greenness" of their products in the consumer market and the use of EPEAT in that market is rapidly growing.

In the institutional market, the attractiveness of a brand is not very important. Professional purchasers rely on specifications and contract language rather than a catchy name and a crisp logo. However, in the consumer market brand image is vitally important. In response to the demand from both environmental and manufacturer stakeholders that EPEAT become a better resource in the consumer market, in early 2009 EPEAT started a project to investigate branding for the consumer market. We have been incredibly lucky to receive the services of one of the world largest and best regarded brand development agencies pro bono to assist us in that effort. We expect to complete that project by YE 2009 with an identity optimized for the consumer electronics market. Our challenge then will be to introduce the new brand to the consumer market.

ENERGY STAR has achieved a remarkable 75% recognition in the consumer market through 20 years of consumer market promotion and brand building, enabled by stable funding provided by a congressional line item. Without the resources to promote EPEAT's new brand in the market it is unclear how fast consumer recognition will grow.

EPEAT COLLABORATION WITH ENERGY STAR

In order to reduce duplication of effort and harmonize with other rstandards worldwide, EPEAT stakeholders incorporated numerous other environmental standards by reference, including the EU RoHS, EPA's Plug Into E-cycling Guidelines, and ENERGY STAR. ENERGY STAR staff participated in the development of EPEAT, and stakeholders have relied extensively on ENERGY STAR product definitions, technical approaches, and energy specifications in developing the energy efficiency requirements of the IEEE/E PEAT standards. As a result EPEAT's primary requirement for energy efficiency is compliance with current ENERGY STAR requirements. Therefore, **all EPEAT registered products meet ENERGY STAR energy specifications.** Stakeholders now developing energy efficiency criteria for EPEAT Imaging Equipment and TV standards continue to use ENERGY STAR specs as the required baseline for those products. In addition, we expect that stakeholders will likely include "more efficient than ENERGY STAR" criteria in those and future standards as part of the optional criteria.

In the past EPEAT has largely relied on ENERGY STAR to verify that ENERGY STAR qualified products actually meet ENERGY STAR criteria. We have discussed with ENERGY STAR the possibility of cooperating on those verifications and expect that this may come to fruition in late 2009 or 2010.

Finally, EPEAT and ENERGY STAR have discussed several options for working together more closely to promote green electronics in the consumer market. It has been difficult to work out the brand/label issues and we continue to discuss options.

In: Environmental Considerations in Federal Procurement
Editor: John G. Kaspin

ISBN: 978-1-61209-532-5
© 2011 Nova Science Publishers, Inc.

Chapter 8

PROCURING SOLAR ENERGY: A GUIDE FOR FEDERAL FACILITY DECISION MAKERS

United States Department of Energy

ACKNOWLEDGMENTS

A special thank you to Charles Hemmeline (U.S. Department of Energy [DOE]) and Anne Crawley (DOE Federal Energy Management Program [FEMP]) for their material support, advice, and feedback.

For contributing time, expertise, and content, the authors thank Kevin DeGroat (Antares), Byron Stafford (National Renewable Energy Laboratory [NREL]), and John Nangle (NREL). For providing information, consultation, advice, and feedback, we thank Kate Anderson (NREL), John Barnett (NREL), Bill Brunner (Straightedge), Nate Butler (Naval Facilities Engineering Command Southwest [NAVFAC SW]), Michael Callahan (NREL), Jason Coughlin (NREL), Douglas Dahle (NREL), Jeffrey Dominick (Johnson Controls), Martha Duggan (Uni-Solar), Mark Frickel (Sentech), Andrea Kincaid (Defense Logistics Agency [DLA Energy]), Dustin Knutson (NREL), Ann Kosmal (U.S. General Services Administration [GSA]), Boyan Kovacic (FEMP), Mark Levi (GSA), Kevin Lynn (DOE), David Marciniak (GSA), Roman Piaskoski (GSA), Doug Porter (GSA), Kevin Prince (SunPower), Scott Provinse (SunEdison), Charles Rienhardt (GSA), Chandra Shah (NREL), Brad Sharp (SunPower), Wayne Taylor (Naval Air Systems Command [NAVAIR]), Karen Thomas (NREL), Sarah Truitt (Sentech), Otto VanGeet (NREL), Andy Walker (NREL), Martin Weiland (GSA), and Bob Westby (NREL).

For comprehensive editing, publication, and Web support, we are grateful to Kathryn Ruckman (NREL), Linh Truong (NREL), Gabe Boeckman (NREL), Christina Van Vleck (Sentech), Susan Carollo (Sentech), and Lisa Joanis Teman (All the Right Words).

The DOE Solar Energy Technologies Program (SETP) and Federal Energy Management Program (FEMP) provided the funding for this chapter.

ACRONYMS AND ABBREVIATIONS

AC	alternating current electricity
ACHP	Advisory Council on Historic Preservation
AFO	alternative financing options
AWC	area-wide contract
BAFO	best and final offer
BLM	Bureau of Land Management
BOA	basic ordering agreement
Btu	British thermal unit
CEC	California Energy Commission
CEQ	Council on Environmental Quality
CFR	Code of Federal Regulations
CO	contracting officer
COR	contracting officer's representative
COTS	commercially available off-the-shelf
DC	direct current electricity
DESC	Defense Energy Support Center (now DLA Energy)
DG	distributed generation
DLA Energy	Defense Logistics Agency (formerly DESC)
DOD	United States Department of Defense
DOE	U.S. Department of Energy
DSIRE	Database of State Incentives for Renewable Energy
EA	environmental assessment
ECM	energy conservation measure
EE	energy efficiency
EERE	DOE Office of Energy Efficiency and Renewable Energy
EIA	Energy Information Administration
EIS	environmental impact statement
EISA 2007	Energy Independence and Security Act of 2007
E.O.	executive order
EPA	United States Environmental Protection Agency
EPAct 2005	Energy Policy Act of 2005
ESA	energy services agreement
ESCO	energy services company
ESPC	energy savings performance contract
EUL	enhanced use lease
FAR	Federal Acquisitions Regulation
FEMP	DOE Federal Energy Management Program
FFS	federal financing specialist
FMV	fair market value
FONSI	finding of no significant impact
FPR	final proposal revision
GHG	greenhouse gas
GSA	U.S. General Services Administration

IDIQ	indefinite delivery, indefinite quantity
IGA	investment grade audit
IMBY	In My Backyard
IRR	internal rate of return
ITC	investment tax credit
J&A	justification and approval
kV	kilovolt
kW	kilowatt
kWh	kilowatt-hours
L&MP	lease and management plan
LCC	life-cycle cost
LEED	Leadership in Energy and Environmental Design
LD	liquidated damages
LPTA	low price, technically acceptable
M&V	measurement and verification
MAC	multiple-award contract
MW	megawatt
MWh	megawatt-hours
NAFTA	North American Free Trade Agreement
NAVAIR	Naval Air Systems Command
NAVFAC SW	Naval Facilities Engineering Command Southwest
NEPA	National Environmental Policy Act
NHPA	National Historic Preservation Act
NMCI	United States Navy/Marine Corps Intranet
NOL	notice of opportunity to lease
NPV	net present value
O&M	operations and maintenance
OMB	Office of Management and Budget
PA	preliminary assessment
PEIS	programmatic environmental impact statement
PF	project facilitator
PO	purchase order
PPA	power purchase agreement
PTC	production tax credit
PV	photovoltaic
PVA	photovoltaic array
RE	renewable energy
REC	renewable energy credit
RFI	request for information
RFP	request for proposal
RFQ	request for qualifications
RPS	renewable portfolio standard
SETP	DOE Solar Energy Technologies Program
SHC	solar heating and cooling
SIR	saving investment ratio
SOW	scope of work

SPB	simple payback
SRCC	Solar Rating and Certification Council
SREC	solar renewable energy credit
SSO	senior sustainability officer
SWH	solar water heating
UESC	utility energy services contract
URESC	utility renewable electricity service contract
Western	Western Area Power Administration
WRL	work release letter

EXECUTIVE SUMMARY

Federal agency leaders are expressing growing interest in generating solar energy on their sites, motivated by both executive branch sustainability targets and a desire to lead by example in the transition to a clean energy economy. *Procuring Solar Energy: A Guide for Federal Facility Decision Makers* presents an overview of the solar project process in a concise, easy-tounderstand, step-by-step format. This guide is designed to help agency leaders turn their interest in solar into successful installations. Detailed information and sample documents for specific tasks are referenced with Web links or included in the appendixes. This guide concentrates on distributed solar generation and does not consider all the complexities of large, centralized solar energy generation.

Because solar energy technologies are relatively new, their deployment poses unique challenges. In addition, solar technologies are commonly used for on-site applications, meaning that a solar project is likely to be connected to both an electrical system and a building. The methods for procuring solar energy are evolving rapidly, ranging from direct purchases to more sophisticated long-term agreements. This guide outlines methods that federal agencies have used to procure solar energy with little to no capital costs. Each agency and division, however, can have its own set of procurement procedures. As a result, this guide was written as a starting point, and it defers to the reader's set of rules if differences exist.

The global solar industry is maturing rapidly, and solar project developers are gaining experience in working with federal agencies. Technology improvements, cost reductions, and experienced project developers are making solar projects easier to put into service. In this environment, federal decision makers can focus on being smart buyers of solar energy instead of attempting to become experts in solar technology.

Federal agencies have many reasons to consider implementing solar energy on their sites, including legislation, executive orders, and agency targets, among others. For agencies who want to pursue solar, the recommended first step is to determine your preliminary solar potential by conducting a solar site screening or an agency-wide screening that identifies the best solar project sites.

A two-part process has been developed as a guide to implementing a smooth and successful solar project. Part 1 of the process includes five project planning steps that cover identification of needs and goals, assembling an on-site team, evaluating the site's solar screening, project requirements and recommendations, and making a financing and contracting decision. Part 2 of the process includes process guidance on the following financing and contracting options:

- **Agency-funded project** – funds have been designated for the outright purchase of a project
- **Power purchase agreement** – a private entity installs, owns, operates, and maintains customer-sited solar energy, and the site purchases electricity or thermal energy through a long-term contract with specified energy prices
- **Energy savings performance contract** – an energy services company incurs the cost of implementing an energy project and is paid from the operations savings resulting from the project
- **Utility energy services contract** – an agreement with a "serving" utility to finance and install an energy project
- **Enhanced use lease** – prospective developers compete for an energy project site lease with payment being either monetary or in-kind consideration; renewable power can be part of the consideration.

The U.S. Department of Energy (DOE) Federal Energy Management Program (FEMP) can provide assistance or answer questions about solar or energy efficiency projects. Call 877-EERE-INF (877-337-3463) or visit www.eere.energy.gov/informationcenter/.

INTRODUCTION

The global solar industry is maturing rapidly, and solar project developers are gaining experience working with federal agencies. Technology improvements, cost reductions, and experienced project developers are making solar projects easier to put into service. In this environment, federal decision makers can focus on being smart buyers of solar energy instead of attempting to become experts in solar technology.

This step-by-step manual guides readers through the process of implementing an on-site solar project. After discussing the importance of solar projects, agency targets, and preliminary solar site screenings at the agency level, the guide presents a two-part process for implementing a solar project on a federal site. Each part has several sub-steps and considerations.

At the point in the process when a financing and contracting decision has been made (see Part 1, Step 5), the process diverges. Part 2 follows the process for each financing and contracting option through to the end of the project. This guide presents the major components of successful solar projects so that readers can achieve their own solar energy goals, which might include a single facility installation, a multi-facility procurement, or an agency-wide procurement plan, among others.

Supplemental information for this guide can be found in the following appendixes: Appendix A presents background material on solar technologies; Appendix B contains a step-by-step self-guided solar screening, along with several other types of checklists; and Appendix C includes case studies of solar projects at federal facilities.

Although this guide focuses on solar projects, it is important to note that energy efficiency is important for reducing both site energy costs and the environmental impacts of using conventional methods of energy production.

Of particular importance is a solar project champion who can help clear project barriers and see the project through to completion. This champion can come from many areas, such as facility or regional management, headquarters leadership, or another leadership area. But no matter where a champion is found, he or she must help move the project forward and overcome barriers that might otherwise stop or delay a project. Experience has shown that strong project leadership is a common factor in successful solar projects.

As the nation's largest energy consumer, the federal government presents a tremendous opportunity for jump-starting a significant increase in domestic solar production. The following subsections summarize some of the reasons why.

The U.S. Department of Energy (DOE) Federal Energy Management Program (FEMP) can provide assistance or answer questions about solar or energy efficiency projects.

Call 877-EERE-INF (877-337-3463) or visit www.eere.energy for more information.

MAKING THE CASE FOR FEDERAL SOLAR ENERGY PROJECTS

As the nation's largest energy consumer, the federal government presents a tremendous opportunity for jump-starting a significant increase in domestic solar production. The following subsections summarize some of the reasons why.

Federal Requirements— Bringing the Backdrop to the Forefront

The Energy Policy Act of 2005 (EPAct 2005) set the primary renewable energy requirements for federal agencies. It requires that renewable energy be tapped—to the extent that is economically feasible and technically practicable—to generate the following percentages of the federal government's total electricity usage:

- Not less than 3% in fiscal years 2007 through 2009
- Not less than 5% in fiscal years 2010 through 2012
- Not less than 7.5% in fiscal year 2013 and thereafter.

EPAct 2005 defines renewable energy as "electric energy generated from solar, wind, biomass, landfill gas, ocean (including tidal, wave, current, and thermal), geothermal, municipal solid waste, or new hydroelectric generation capacity achieved from increased efficiency or additions of new capacity at an existing hydroelectric project" (www.eere.energy.gov/ femp/technologies/renewable_requirements. html; accessed June 8, 2010). By granting a bonus for electricity from renewable energy generated on a federal or tribal site, EPAct 2005 also created a strong incentive for on-site projects. Executive Order (E.O.) 13423, Strengthening Federal Environmental, Energy, and Transportation Management, signed on January 24, 2007, sets even more challenging goals for the federal government. The 2007 E.O. mandates that 50% of all renewable energy required under EPAct

2005 must come from "new" renewable energy (meaning energy from facilities placed in service after January 1, 1999). The guidance under E.O. 13423 allows a limited amount of thermal energy to count toward the E.O. goal for new renewable energy, but not toward the EPAct goal. (See www.eere.energy.gov/femp/ regulations/eo13423.html.) The FEMP guidance on E.O. 13423 and EPAct 2005 requires agencies to own the renewable energy credits (RECs) associated with any renewable energy counted toward the goal.

The Energy Independence and Security Act of 2007 (EISA 2007) requires that 30% of the hot water demand of new federal buildings (and major renovations) be met with solar water heating equipment, as long as the solar system remains cost effective over its life cycle (www.eere.energy. html; accessed July 20, 2010).

E.O. 13514, Federal Leadership in Environmental, Energy, and Economic Performance, was signed on October 5, 2009. This order establishes "an integrated strategy towards sustainability in the Federal Government" and makes "reduction of greenhouse gas emissions a priority for federal agencies" (www.eere.energy eo13514.pdf; accessed June 8, 2010). This E.O. builds on the federal energy efficiency mandates of EPAct 2005, EISA 2007, and E.O. 13423 by using greenhouse gas (GHG) emissions as a unifying metric for federal sustainability. The order requires agencies to:

- Appoint a senior sustainability officer (SSO)
- Establish a GHG-emission baseline for fiscal year 2008
- Set GHG-emission reduction targets for fiscal year 2020
- Create a strategic sustainability performance plan to document progress toward achieving the fiscal year 2020 goals
- Inventory and report its GHG emissions for the previous fiscal year (beginning in January 2011).

Each agency is responsible for setting its own goals for GHG emission reduction through a strategic sustainability plan, with review and approval by the chair of the Council on Environmental Quality (CEQ) and in consultation with the director of the Office of Management and Budget (OMB).

Environmental Protection— Doing the Right Thing

In the United States it is estimated that in 2010, energy consumption in buildings will generate 38% of U.S. carbon dioxide emissions. As some of the largest single owners and occupants of buildings, federal agencies have the opportunity to greatly reduce carbon dioxide emissions by implementing energy efficiency and renewable energy measures (2009 Buildings Energy Data Book, table 1.4.1; available at http:// buildingsdatabook .eren.doe.gov/TableView. aspx?table=1.4.1; accessed July 9, 2010).

Photovoltaic (PV) and solar thermal installations emit no GHGs as they operate. By installing a PV system, an agency can generate clean electricity and reduce the amount of electricity it needs to purchase. And installing solar thermal systems can reduce electricity use or the direct use of fossil fuels like natural gas, with a bonus of decreased GHG emissions.

Installing solar energy, then, increases the use of renewable energy and reduces an agency's carbon footprint at the same time.

The U.S. Environmental Protection Agency (EPA) maintains a Web site that lists GHG emission factors for electricity from different regions in the United States (http://cfpub.epa.gov/egridweb/ghg.cfm; accessed July 20, 2010). The GHG emissions vary with the mix of fuels utilities use to generate electricity. Because using solar energy displaces peak plant production, the non-baseload emission factors are the most appropriate when estimating GHG emission reductions. Most solar thermal projects do not require separate calculations of GHG emission reductions because they directly reduce the use of fossil fuel or electricity that a site would otherwise have to report in its GHG inventory.

DOE Office of Energy Efficiency and Renewable Energy (EERE) Web sites offer a wealth of information about the various requirements. See, for example, the following:

Federal requirements are listed at www.eere.energy renewable_requirements.html.

FEMP training courses for federal renewable energy goals can be found at www.eere.energy.gov/femp/technologies/ renewable_training.html.

Additional resources are available at www.eere.energy renewable_energy.html and www.eere.energy.gov/femp/technologies/ renewable_solar html.

Guidance on federal facility reporting requirements is available at www.eere.energy facility_requirements.html.

When it is final, guidance on GHG accounting and reporting under E.O. 13514 will be accessible on the FEMP Web site.

President Barack Obama, Col. Dave Belote, 99th Air Base Wing commander, and Senator Harry Reid, D-Nevada, tour the Nellis Air Force Base, Nevada, photovoltaic array (PVA) on May 27, 2009. The PVA has over 72,000 solar panels which save the base approximately $1 million a year in energy costs. (*Photo Credit: Senior Airman Nadine Y. Barclary, Nellis Air Force Base, PIX 17160*)

Leadership by Example— Paving the Way

As the largest consumer of energy and, therefore, the greatest emitter of GHG in the U.S., the federal government can greatly impact GHG reduction and accelerate the solar market by implementing solar energy projects. Federal agencies have many reasons to lead by example and to provide visible demonstrations of solar projects. E.O. 13514 sets the vision for federal agency management to lead by example to help "create a clean energy economy that will increase our Nation's prosperity, promote energy security, protect the interests of taxpayers, and safeguard the health of the environment" (www.eere. energy.gov/femp/pdfs/eo13514.pdf, accessed June 8, 2010). Federal agencies will reach this vision by reducing GHG emissions through building or retrofitting more efficient buildings, increasing the use of renewable energy, and reducing fleet petroleum use.

Designing or renovating a facility to Leadership in Energy and Environmental Design (LEED) standards is also part of leading by example. Including a solar project in any design or retrofit helps earn points toward LEED certification. Typically, the number of points a facility can earn toward a LEED rating is based on the percentage of building energy cost that is offset by the system. More information about how renewable energy can affect a LEED project can be found at www.leeduser.com/leed-credits (accessed July 2010); choose the appropriate LEED rating system and then select the "Energy and Atmosphere" category for renewable energy.

Part of leading by example and meeting an agency's renewable energy targets is documenting the installed system's performance. The DOE Solar Energy Technologies Program (SETP) tracks the performance and reliability of system installations. For more information about having DOE track a specific system, contact Michael Quintana at Sandia National Laboratories (maquint@sandia.gov).

Cost Savings—Doing the Math

Solar technology advances and improved financing approaches are steadily reducing the cost of implementing solar energy projects. In many cases, solar energy is the lower cost energy solution. As an added benefit, it's possible to know the energy costs for the system's output for 20 or more years. The cost of energy from fossil fuel production can be volatile, and with the potential need to account for climate change and GHG emissions in economic terms, the benefits of a currently competitive and known cost of energy are clear. When a good site, supportive incentives, and strong leadership come together, solar becomes a smart investment.

If site conditions preclude implementing a solar project, an agency can still purchase renewable energy. More information is available at www.eere.energy.gov/femp/technologies/renewable_purchasingpower.html (accessed June 8, 2010).

Agency-wide Planning: Solar Energy Screening

One of the first steps agencies can take to begin pursuing solar is to conduct an agency-wide solar energy screening. An agency-wide solar energy screening is a preliminary assessment of the viability of the agency's sites to host solar energy systems. This screening will include a basic overview of the potential of sites to host solar installations with approximate size estimation and energy production. The screening process enables agency decision makers to select the best sites for renewable energy investment and to meet their renewable targets more economically. In some cases, a DOE laboratory can perform the screening. DOE labs have several resources to help with screenings such as NREL's Renewable Energy Optimization (REO) service. A solar screening can also be conducted internally or externally under an energy savings performance contract (ESPC) or by engaging a contractor listed in the U.S. General Services Administration (GSA) Comprehensive Professional Energy Services Blanket Purchase Agreement (available at www.gsa.gov/portal/content/105339; accessed July 19, 2010). The organization selected to perform the agency-wide screening must have a good understanding of solar technology applications and performance. A successful solar screening will give agencies valuable information they can use to begin the process for developing targeted site or regional solar procurements, described in the next section of this guide. More information related to solar screening can be found in Appendix B on page 77.

PART 1. PLAN A SOLAR ENERGY PROJECT

The impetus for implementing a specific solar energy project may come from top-down agency guidance such as a solar energy screening, or from specific site needs and goals. This section walks through the steps needed to execute a specific site solar procurement and is intended to help those charged with carrying out the solar project. Though this section largely considers single site installations, agencies should look for innovative ways to aggregate procurements as much as possible to benefit from economies of scale and to reduce transaction costs.

THE PROJECT PLANNING PROCESS

Details on each step are included in this section

Step 1. Identify needs and goals
Step 2. Assemble an on-site team
Step 3. Evaluate candidate solar energy sites
Step 4. Consider project requirements and recommendations
Step 5. Make a financing and contracting decision

Step 1. Identify Needs and Goals

Several common reasons for considering a solar project follow:

- The agency must meet renewable energy targets.
- The appropriations are available for improving a facility.
- The project is a good way to meet a site's needs—depending on site conditions, there can be many smart reasons to implement a solar project.
- The project can provide energy cost savings.
- The project can reduce future energy cost volatility and uncertainty.
- The project will earn credits toward LEED certification.

The reasons for considering a solar project help to define the needs and goals that the project will address. Needs and goals comprise the vision, the touchstone, and the principles that guide the process of setting priorities, creating decision criteria, and making decisions. Solar should be part of a broader vision of whole systems design for buildings and sites. Potential goals or criteria include the following:

- Maximize on-site solar energy production (particularly within a restricted budget).
- Maximize the return on investment.
- Meet a minimum annual solar energy production target.
- Maximize GHG reductions.

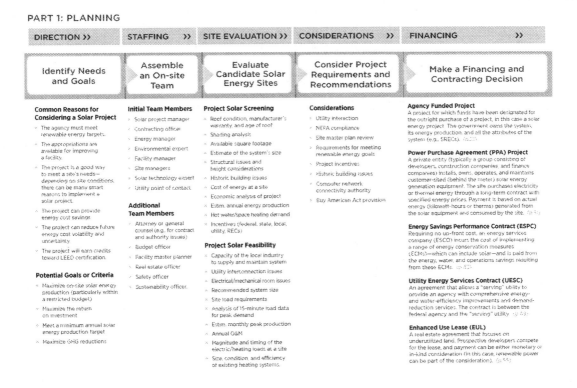

Goals could adjust or change as the project develops, but they always should be at the forefront during the decision-making process.

Step 2. Assemble an On-Site Team

At this point, a solar project team should be identified. The team is important not only for getting the work done, but also for making sure that all issues are considered. Even small oversights can be costly in terms of dollars and time, and can even result in a failure to accomplish project goals.

One of the most important features of the team should be its alignment with the project's goals. The project goals can adjust with team input— and healthy debate on project questions is useful—but if any team member hasn't bought into the goals *before* being invited to join the team the project won't go smoothly. Referring to Step 4, "Consider Project Requirements and Recommendations," can help when considering the makeup of the team. That step outlines the diverse considerations that feed into successful project completion. It's important to recognize that it takes a diverse group of people with a wide range of skills to bring a project to fruition.

The initial solar project team might be small and include only those members relevant to the immediate task; this type of team can grow as the project requires. As an alternative, the team could include—from its inception—everyone who has a stake in the project process. This decision should be based on best judgment and staff availability.

PV solar array installed at Marina Del Rey USPS Postal Station in Los Angeles, California. (*Credit: George Marsh, PIX 11015*)

If starting with a small team, the people who should participate, particularly in Step 3, where the site's solar screening step is evaluated, include the following:

- Solar project manager
- Contracting officer
- Energy manager
- Environmental expert (responsible for environmental review under the National Environmental Policy Act [NEPA])
- Facility manager
- Site managers (if multiple people are responsible for different parts of the site)
- Solar technology expert (depending on procurement option selected)
- Utility point of contact.

As the project progresses, adding other team members should be considered:

- Attorney or general counsel (e.g., for contract and authority issues)
- Budget officer
- Facility master planner
- Real estate officer
- Safety officer
- Sustainability officer.

Several factors will ensure the team's success. For example, a high-level person from the agency should participate. This person doesn't need to be involved in the details of the project, but should check in periodically to help move the project forward and overcome barriers that might otherwise stop or delay a project. It's also critical to select a contracting officer and legal advisor with strong leadership characteristics, because it might take initiative and innovation to push a project through ambiguous areas of the procurement process. Team dedication and creativity are crucial as well, because these traits are essential for finding innovative, cost- effective solutions, if necessary. The solar energy system procurement process is relatively new, and although challenges are being addressed and resolved, issues may still exist. Navigating around these obstacles requires leadership, commitment, and creativity.

Building and site occupants are also stakeholders in the project. A representative of this group can be included in the team as a liaison to keep communications open— especially if the project will have any sort of significant impact on occupants.

After the team is assembled, its roles, responsibilities, and timelines should be established. Scheduling periodic meetings will keep the project moving forward on track.

Step 3. Evaluate Candidate Solar Energy Sites

This section outlines the detailed information needed to evaluate specific sites for solar energy installations. These findings may be available from an agency-wide solar screening if your agency has one, or they may need to be compiled by the solar project team.

United States Coast Guard – The Williams Building in downtown Boston, Massachusetts, now has a 28 kW AC PV system integrated into the roof consisting of 372 panels. (*Credit: SunPower, PIX 08466*)

There are two recommended levels of solar site evaluation:

1. A project solar *screening*, which is a high-level, preliminary analysis used to determine a site's likely viability, and
2. A project solar *feasibility study*, which is a more rigorous engineering and economic analysis to define specific system design considerations for use in requests for proposals and/or scope of work development.

For projects that propose to use alternative financing (see Part 2), a project solar *screening* is sufficient to proceed. For agency funded projects, a solar *feasibility study* is recommended. The points that both types of solar evaluations should cover are defined in the sections that follow.

If the team finds that the site screening doesn't cover all the issues relevant to the project and site, a more complete feasibility study should be obtained. To determine what resources might be available based on the specifics of a project, call FEMP (202-586-5772) or hire a private contractor to perform the study. Be sure to address all the relevant points listed below based on the individual site. If the screening or feasibility study covers all of the relevant issues but the quality of the report is questionable, a first-order check can be performed using the solar screening evaluation checklist located in Appendix B.

Project Solar Screening

A project solar screening should encompass the following.

Structural or Mechanical Considerations

- Roof condition, manufacturer's warranty, and age of roof (if considering a rooftop system)
- Shading analysis (identification of obstructions that might shade the array location)

- Available square footage for a solar system
- Preliminary estimate of the system's size
- Structural issues and—if the system is to be mounted on a building—height considerations
- Historic building issues (if the system will be on a building that could be a historic property or is located in a historic district).

Economic Considerations

- Cost of energy (electricity and fuels) at a site, plus any details of rate schedules that could favor or penalize solar
- Economic analysis of project (e.g., simple payback, internal rate of return [IRR], net present value [NPV], life cycle cost [LCC], projected savings)
- Estimated annual energy production
- Hot water or space heating demand
- Incentives (federal, state, local, utility, RECs) and their time sensitivities.

Project Solar Feasibility Study

A project solar feasibility study should encompass the following in addition to the project solar screening components.

Structural or Mechanical Considerations

- Capacity of the local industry to supply and maintain such systems
- Utility interconnection issues (if planning an electric project, it's important to know whether the utility has special hardware or contractual requirements)
- Electrical room or mechanical room issues (e.g., space for equipment, alternate location, capacity limits, access between system and equipment room).

Alfred A. Arraj U.S. Courthouse; Denver Federal Courthouse; PV system supplied by Atlantis Energy Systems PV shade structure on the top floor is cantilevered over facade; first federal courthouse to get PV glazing. (*Credit:Atlantis Energy Systems, Inc., PIX 13999*)

Economic Considerations

- Recommended system size
- Site load requirements (these should be checked against system sizing)
- Analysis of 15-minute load data for peak demand
- Estimated monthly peak production
- Annual operations and maintenance (tasks, annual costs)
- Magnitude and timing of the electric and heating loads at a site
- Size, condition, and efficiency of existing heating systems.

Step 4. Consider Project Requirements and Recommendations

If, at this point, the solar screening demonstrated that the project is viable, the following should be considered:

- Utility interaction
- NEPA compliance
- Site master plan review
- Requirements for meeting renewable energy goals
- Project incentives
- Historic building issues
- Computer network connectivity authority
- Buy American Act provision

Utility Interaction

If a project includes PV, it also includes an electrical interconnection with the utility. The interconnection agreement is made between the organization and the utility regardless of the solar project developer's role. It's important to communicate with the utility about the proposed project early in the process. This ensures that all interconnection issues are taken into account early on, and helps avoid unpleasant surprises after significant effort has been expended. Depending on the utility and the local distribution system design, the project can be adversely affected by expensive interconnection hardware and requirements. Some requirements can be addressed during the design stage through equipment specifications, which incurs only a nominal extra cost as compared to purchasing additional equipment.

Changes in the electricity tariff rate structure should also be explored and discussed with the utility. Based on the system's projected hourly and seasonal performance, a more optimal tariff could be available that could potentially offset both energy (kilowatt-hour [kWh]) charges and demand (kilowatt [kW]) charges. Although utilities are important partners in the process, they might not be enthusiastic about reduced electrical consumption. Thousands of utilities operate in the United States—each with unique rate structures and policies—so it's important for those planning a renewable energy project to understand their rights as consumers and know what impact the solar system will have on future utility billings. For example, a utility could impose a standby charge to cover the cost of maintaining generation resources that are used when the solar energy system is not generating. An agency should also determine whether it can sign a utility interconnection agreement that has indemnification clauses.

The system owner and utility will eventually develop an interconnection agreement that defines all the specific requirements and terms of the interconnection. Information on state-specific interconnection standards can be found at www.irecusa.org/index.php?id=86 (accessed June 8, 2010).

NEPA Compliance

Opinions differ among agency representatives about when NEPA should be addressed during the process, but it's best to consult the environmental expert responsible for NEPA early in the process. This will ensure that the expert is informed about the decisions and the directions regarding the site locations for the project. This (and possibly other information) will help guide information that will eventually go into the request for proposal (RFP) and ensure that the project stays on track. The following is a general overview of the National Environmental Policy Act.

Trigger

If the project is located on federal land or uses federal funding, it must comply with NEPA. The effort involved to comply with NEPA greatly depends on where the project is located on the site and also on the project's scale. The impact of a rooftop system, for example, is typically less than that of a ground-mounted system. As a result, it's usually easier for such systems to comply with NEPA. A NEPA resource can be found at www.epa.gov/Compliance/ resources/ faqs/nepa/index.html (accessed June 8, 2010).

By using solar power to generate electricity, agencies can reduce GHG emissions and produce environmental attributes from the solar power project (e.g., RECs)

Process

The agency environmental expert assesses the proposed system, and the assessment produces one of the following three results:

Level 1 – Categorical exclusion

A letter stating exclusion is issued and the process is complete. (Note that rooftop and small ground-mounted projects generally receive a categorical exclusion.)

Level 2 – Environmental assessment (EA)

This has two possible outcomes.

- A. Finding of no significant impact (FONSI): If an EA is required, the best outcome is receipt of a FONSI. This process usually costs less than $500,000 and takes less than 6 months.

B. If the EA finds that environmental consequences could be significant, an environmental impact statement (EIS) must be completed.

Level 3 – EIS

If the environmental expert finds that the project could have a significant environmental impact from the start, an EIS is completed without completing an EA first. This process can cost $1 to $2 million and take up to 18 months. If successful, a record of decision is received and the project can move forward.

Site Master Plan Review

If there is a master plan for the site, it should be reviewed at this stage. A solar energy project is a long-term commitment, typically in place for 25 years or longer. When reviewing a master plan from a 25-year perspective, things to consider include plans for undeveloped land that might be a site for ground-mounted collectors or, in the case of rooftop arrays, determining if and when a building is scheduled for retirement. Also important is whether any architectural plans include aesthetic features that could preclude the installation of solar energy equipment. In such cases, site managers have found that their project site options can be significantly limited.

Requirements for Meeting Renewable Energy Goals

Meeting federal renewable energy goals is a common motive for implementing a renewable energy project. When this is the case, it's important to understand the rules for counting renewable energy toward the EPAct 2005 requirements and the E.O. 13514 GHG requirements, as well as toward individual agency mandates. By using solar power to generate electricity, agencies can reduce GHG emissions and produce environmental attributes from the solar power project (e.g., RECs). This includes all environmental attributes under federal guidance under EPAct 2005 and E.O. 13423. The agency can choose to sell these attributes, allow the developer to sell them, report them to meet EPAct 2005 requirements, or use them to reduce carbon dioxide emissions under E.O. 13514. Under generally accepted GHG accounting principles, however, an entity must own the RECs from a project before it can claim emissions reductions. The rules governing REC usage vary from agency to agency, they aren't always intuitive, and they dictate how project generated RECs are used. The agency energy manager can explain the agency's approach.

General Services Administration (GSA) gives an example in which more valuable PV RECs were sold to a local utility for $0.24/kWh and inexpensive replacement RECs from wind energy were purchased for less than $0.01/kWh. This greatly increased the economic viability of the PV project. Based on FEMP guidance for the EPAct renewable energy goal, because the renewable energy project is located on federal land, this transaction enabled the party to claim the bonus for the electrical production from the project. The FEMP allows substitutions of RECs for projects on federal or tribal lands to claim the bonus under the EPAct 2005. The generally accepted rules for GHG reporting are different, allowing only single counting of the environmental attributes owned (in this case, the purchased wind energy RECs). Federal guidance for GHG accounting under E.O. 13514 is currently being reviewed. Visit the FEMP Web site for current information (www.eere.energy. gov/femp/; accessed June 8, 2010).

One problem with selling more valuable solar RECs and buying cheaper replacement RECs is the uncertainty surrounding the ability of federal agencies to sell federal property (in this case the solar REC). Solutions that will allow other agencies to sell solar RECs are being evaluated. One solution, similar to the approaches used to take advantage of incentives, is to arrange for the private developer to own and sell the RECs in return for offering the agency a better deal on the project. In this case, because the agency never takes ownership of the RECs, it's not directly involved in the sale.

Project Incentives

The solar screening should include all potential incentives that would help the economics of the project. At this point in the process, it's wise to ensure that all incentives have been included and important to understand federal requirements and the agency's policies on incentives. The economics of renewable energy projects are often dependent on federal and state incentives, and the federal government is ineligible for many of these incentives. Private developers, however, can take advantage of the tax credits, grants, and other incentives that drive the renewable energy markets. This becomes a key consideration in deciding whether to fund the solar project through direct appropriations or alternative financing.

Based on FEMP guidance for the EPAct renewable energy goal, because the renewable energy project is located on federal land, this transaction enabled the party to claim the bonus for the electrical production from the project.

Most states have energy incentive programs that help offset energy costs and promote energy efficiency and renewable energy technologies. FEMP staff members research these programs on a state-by-state basis to help federal agencies meet their energy management goals. (See www.eere.energy.gov/femp/financing/ energyincentiveprograms.html; accessed June 8, 2010.) A Database of State Incentives for Renewable Energy (DSIRE) is available at www.dsireusa.org (accessed June 8, 2010). A summary and details for incentive programs can be accessed by state, and information on federal incentives is available as well. This information can aid in determining which incentives apply to an agency's specific circumstances and help estimate any economic impact on the project. It is also critical to consider the time limitations on the incentives and confirm that they will still be offered when the project will be applying to receive them. Although DSIRE information is typically up to date, it's always a good idea to verify the status and availability of incentives with the administering agency or utility.

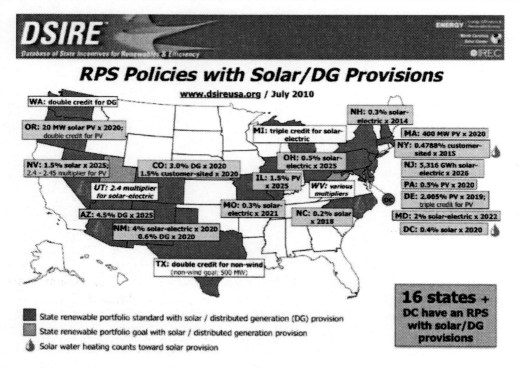

The DSIRE map below illustrates the Renewable Portfolio Standards (RPS) across the United States that also include a solar or distributed generation carve out.

Renewable energy credits are the property right attribute created when electricity is generated by a renewable energy source. The RECs specific to solar energy generation (solar renewable energy credits or SRECs) generally have greater value. The RECs and SRECs can be sold to a utility to help meet its renewable portfolio standard (RPS) needs, sold on the voluntary market, or "retired" and counted by the agency toward its mandated renewable energy requirements. If RECs are sold or retained by the developer, however, the project no longer qualifies as renewable energy required to meet the EPAct 2005 goal and likely will not be allowable for reducing GHG emissions.

An agency might also have a policy on accepting certain incentives. Several agencies are evaluating whether accepting financial incentives from limited financial pools reduces the incentives available to private-sector investment. If this is the case, agencies could establish policies prohibiting the use of available renewable energy and energy efficiency incentives in projects.

Historic Building Issues

Section 106 of the National Historic Preservation Act (NHPA; www.achp.gov/nhpa.html; accessed June 8, 2010) requires federal agencies to consider the effects of their projects (or projects requiring their assistance or approval) on historic properties, and to give the Advisory Council on Historic Preservation (ACHP) an opportunity to comment. Historic properties are defined as "any prehistoric or historic district, site, building, structure, or object included in or eligible for inclusion in the National Register, including artifacts, records, and material remains related to such a property or resource" [16 U.S.C. § 470(w)]. Section 106 of the NHPA does not mandate preservation, but outlines a process for considering alternatives and allows the public to influence decision making. For more information, see www.achp.gov/ docs/Section106FactSheet.pdf (accessed June 8, 2010).

If a project can cause adverse effects on historic sites or buildings, a legally binding agreement is typically used to resolve those effects. If no agreement can be reached, the ACHP issues advisory comments that the head of the agency must consider. Regulations governing the review process are found at www.achp.gov/ regs-rev04.pdf (accessed June 8, 2010).

Computer Network Connectivity Authority

Many renewable energy systems—especially electric systems—require automated monitoring and control. This is usually accomplished by connecting the renewable energy system to a facility's existing building-monitoring system through a computer network connection. Some computer networks are operated under contracts that have very specific requirements. These contracts can be restrictive, allowing only network connection of specific devices, and sometimes even permitting only a limited set of preapproved software and hardware. It's crucial that the parties controlling the computer network be involved early in the process, because it's difficult and time- consuming to acquire the "authority to operate" a system connected to the network. An example is the United States Navy's computer networks, which are operated by a contractor under the Navy/Marine Corps Intranet (NMCI) contract. A renewable energy system that required a network connection was installed, but because of the contract it couldn't be operated until required permissions were obtained—which was a long and difficult process.

Buy American Act Provision

The Buy American Act restricts the federal government's purchase of supplies and construction materials that are not made domestically. The act contains many provisions, and when looking at the specification of components for a solar system, it's difficult to determine which products comply and which do not. The current understanding is that a product manufacturer self-certifies its products as meeting the provisions of the Buy American Act, and that an audit system exists for this certification process. This guide cannot confirm the correctness of this process, but it provides the best information available at the time of publication. If a project team decides to install solar equipment that complies with the Buy American Act, the equipment can be purchased directly from a manufacturer that meets this requirement, or this requirement can be specified in the RFP and in the final contract. It's important to consult with an agency's buyers and procurement office about their latest understanding of the agency's specific requirements.

Step 5. Make a Financing & Contracting Decision

If a project reaches the financing and contracting point in the process and is moving forward, the solar screening has established— and the project team concurs—that a solar project is worth pursuing through the next step, "Make a Financing and Contracting Decision." Unless funding is designated for the project (i.e., the agency will fund the project), this can be a complex decision. If no direct funding is available, financing options must be considered. Before choosing an available financing option, review the options and information presented in this guide. Then contact a financing specialist to discuss the specifics of the project and confirm the appropriateness of the financing decision. For option-neutral direction on this topic, contact Michael Callahan at National Renewable Energy Laboratory (NREL) (michael.callahan@nrel.gov). In addition, FEMP maintains a Web site and offers Webinars and workshops to educate participants on the different financing options available. The FEMP financing information can be found at www.eere.energy mechanisms.html (accessed July 26, 2010). The FEMP training information is available at www.eere.energy. gov/femp/services. html (accessed June 8, 2010).

Installation of skylight glass laminated with PV cells at the Thoreau Center for Sustainable Development, Presidio National Park, San Francisco, California.. *(Credit: Lawrence Berkeley Lab, PIX 01056)*

The financing options considered in this guide follow:

- Power purchase agreement (PPA)
- ESPC and optional energy services agreement (ESA)
- Utility energy services contract (UESC)
- Enhanced use lease (EUL).

This guide also describes the process for agency-funded or direct-appropriation projects.

PART 2. EXECUTE A SOLAR ENERGY PROJECT

After a financing and contracting option has been selected, the next part is to execute the project. Each project follows a process that is unique to the financing and contracting option selected. The ESPC option has a rigid process and the UESC has a less rigid process. The other processes, however, can be seen more as recommended guidelines. For each of the options listed, the sections below describe general characteristics, provide case studies and project process, and list available resources.

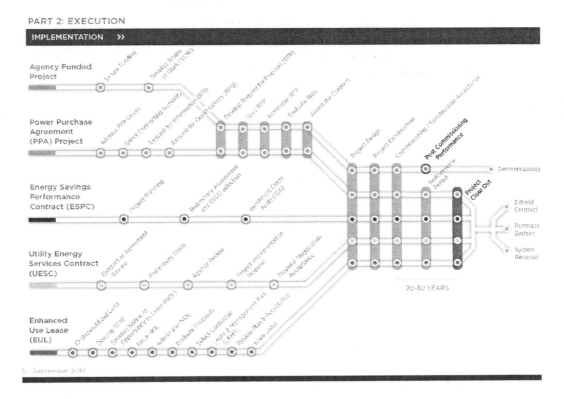

Financing Options

- ☒ Agency Funded Project
- ☐ Power Purchase Agreement (PPA) Project
- ■ Energy Savings Performance Contract (ESPC)
- ☐ Utility Energy Services Contract (UESC)
- ☒ Enhanced Use Lease (EUL)

AGENCY-FUNDED PROJECT

An agency-funded or agency-appropriated project is one for which funds have been designated for the outright purchase of a project, in this case a solar energy project. The government owns the system, its energy production, and all the attributes of the system (e.g., SRECs).

PROS	CONS
• Well-understood mechanism. • Common to many federal capital projects. • Does not incur any financing costs. • Long-term energy-cost reduction.	• Site is responsible for operations and maintenance arrangements (including inverter replacement) but can purchase an operations and maintenance (O&M) service contract. • No assurance of long-term performance (can purchase optional long-term performance guarantees, which differ from a manufacturer's warranty). • Could be more human-resource intensive than other options. • Will not be able to monetize available tax incentives.

Steps to Follow

1. Secure funding
2. Develop the scope of work
3. Develop a request for proposal
4. Issue a request for proposal
5. Administer a request for proposal
6. Evaluate the proposals
7. Award the contract
8. Design the project
9. Construct the project
10. Commission the system
11. Post-commissioning performance

1. Secure Funding

Unless funding has been secured for this project, a case must be made for the project and funding must be requested. The primary source for this funding will be the individual agency—as it is the agency that is responsible for meeting renewable energy targets. Typically, agencies want to examine a business case. The data for a business case should be in the project solar feasibility study but, in some cases, additional data or analysis must be confirmed or put in place. An example is a PV project that was evaluated in 2003 but which did not meet the financial requirements for funding. Three years later (2006), a state law requiring that a percentage of the utility's electricity come from solar power was passed. This essentially required a utility to buy SRECs, which sold for a premium of $0.24 per kWh. Thus, the project changed from an unacceptable investment to a good investment and funding was granted. This is a good example of how incentives can help make the business case for a project.

> **Case Studies**
>
> **(see Appendix C)**
>
> - United States General Services Administration, Federal Center PV Project, Denver, Colorado
>   1.19-megawatt (MW) PV www.gsa.gov/portal/content/105165
> - Social Security Administration, high-temperature hot-water project, Philadelphia, Pennsylvania
>   54-m^2 solar thermal system
>   www.eere.energy energy_projects_detail.cfm/id=8

Note that it is recommended that a project solar feasibility study be completed before this process is started. For more information, see Part 1, Step 3, "Evaluate Candidate Solar Energy Sites."

Social Security Administration, high-temperature hot water project, Philadelphia, Pennsylvania, 54-m2 solar thermal system (*Credit: Ed Hancock, Mountain Energy Partnership*)

2. Develop the Scope of Work

The scope of work becomes the basis of the RFP used to solicit project proposals from solar developers. The form and detail of this scope of work (SOW) are based on the selection of contract type. Typical contract types are:

- GSA schedule,
- Design-bid-build, and
- Design-build.

The GSA schedule identifies contractors that specialize in solar installations. Several of these contractors offer turnkey solutions and can assess the site, design the system, and install the system. Pricing is based on published rates, but vendors might offer discounts. The GSA schedule system can be a quick and efficient contracting mechanism.

Using the design-bid-build method, the scope of work is built around a complete design and specification of the system. This is a very prescriptive approach and the solar developer bids using the complete specifications. Typically this approach is used only when there are very specific design and engineering requirements which necessitate a defined design. In comparison, what is sought in solar projects typically is more performance based.

For the design-build process in the case of a solar project, the scope of work is built around a performance specification. The performance specification typically is not prescriptive and it allows the proposing developers latitude to design systems using their unique competitive advantages. This can lead to good performance per dollar invested.

The project's solar feasibility study should provide the basis for the project's scope of work. The team's solar expert is a key person in the development of this document. The SOW should include:

- Location on-site,
- Performance specification (or project design in the case of design-bid-build),
- Specific site requirements, and
- Type of solar system.

A performance specification can include the following.

- The performance metric used to define the performance. This could include DC output, AC output, and delivered energy. It is not difficult to measure the output of a solar system, but a calculation often is required to normalize these measurements to conditions (solar resource, load) used to derive the performance statement. A performance specification for PV, for example, could state minimum performance based on actual measured solar (e.g., not less than 80% of calculated levels). A specification for solar thermal could state that the system operate above a stated level of efficiency when measured (e.g., 25%).
- Inverter warranty and efficiency (e.g., a minimum of 10 years, 94% California Energy Commission (CEC) weighted efficiency). See, for example, the Go Solar

California "List of Eligible Inverters," at www.gosolarcalifornia.org/equipment inverters.php (accessed June 9, 2010).
- PV module warranty (e.g., 90% of nameplate rating for first 10 years, 80% at 25 years). See, for example, the Go Solar California "List of Eligible SB 1 Guidelines Compliant Photovoltaic Modules," at www.gosolarcalifornia.org/equipment modules .php (accessed June 8, 2010).
- A solar heating- and cooling-system warranty—if it is a custom-built system such as a large concentrator array. For commercial flat-plate and evacuated tube collectors, it is common to specify collectors and systems that are tested to Solar Rating and Certification Council (SRCC) standards and certified by the SRCC. Federal tax credits and most state incentives require SRCC certification. Testing and certification standards currently are being developed for air collectors.
- Minimum annual kWh or Btu production (if needed, for example, to meet minimum SREC production target).
- Interface requirements. Any physical and operational requirements that are imposed by connecting to existing systems (e.g., locations, voltages, temperatures) in such a way that performance of both solar and conventional systems is optimized.
- If a system is ground-mounted, then vegetation preparation and re-establishment as well as its height above the ground should be addressed. This is important because shading due to vegetation and snow accumulation at the lower edge of collectors can impair performance.

Sample performance specifications and SOWs are available at https://www.fbo.gov/download/ d22/d22e91f03f8ff7af7 105083a2b0b7f94/ ARRA _Roof _Final _Specs _9-28-2009. pdf (accessed June 9, 2010); and https://www.fbo.gov/fil es/archive/ade/ ade3 1d57bafa783a1 8f8d802437a060a.pdf (accessed July 19, 2010).

The GSA schedule identifies contractors that specialize in solar installations. Several of these contractors offer turnkey solutions and can assess the site, design the system, and install the system.

AGENCY-FUNDED PROJECT CASE STUDY

Denver Federal Center (GSA)

The Denver Federal Center (DFC) is a 640- acre, secured federal facility operated by the GSA. Twenty-six federal agencies occupy approximately 50 active buildings, with approximately 6,000 employees on site. The GSA is striving to make this facility the most sustainable business park in the country. In 2004, the state of Colorado established a renewable energy standard, which started the process by which the DFC could obtain its first PV electrical generating facility. The REC system created by the standard enabled the DFC to sell RECs produced from a PV system at a price that made the system economical. With technical assistance from NREL, the DFC awarded the design-build contract through a competitive RFP and was able to get the project installed in just a few months to meet REC sale obligations. The system is a 1.19 MW fixed tilt system that generated 1,726 megawatt hours (MWh) in 2008. Generation in 2008 was 2.5% of DFC's total annual load. Less expensive RECs from Texas were purchased to help the GSA meet its renewable energy targets. A performance specification in the contract allowed the DFC, who owns the system, to confirm the continued performance of the system during the first year of operation.

3. Develop a Request for Proposal

An RFP is the document issued to the public to solicit proposals; in this case, from solar developers. (If using the GSA schedule, then an RFP is not required and there is no need to go through the RFP process.) The RFP describes how the proposal process is to be conducted and provides information that can be used as a basis for a developer's proposal. An RFP should include the following elements (listed alphabetically, and not in order of importance).

- Clarification of party responsible for procuring permits
- Commissioning plan
- Criteria and process to be used to evaluate proposals
- Definition of infrastructure requirements (if any)
- Delineation of historic building requirements (if any)
- Description of NEPA requirements (if any)
- Due diligence
- Explanation of how the proposal process is to be administered (e.g., proposal meetings, site visits, responses to questions)
- Limits on proposed project timeline
- Requirements for priced options (for example, extended warranty and maintenance agreements priced in 5-year increments)

- Restrictions (or preferences) on parties allowed to submit proposals (e.g., small business, woman owned, veteran owned); this is a policy decision that can be based on agency- or site-specific preferences
- SOW
- Specification of post-commissioning performance
- Timelines for proposal process.

Criteria and Process for Evaluating Proposals

Describing the criteria and the process to be used to evaluate the proposals helps developers to structure their responses and ensures that the project's primary issues are addressed. This also makes it easier to review proposals. The respondents address identified criteria in separate sections, eliminating the need for a reviewer to pick statements from the entire proposal that apply to the criteria. The three common processes used for evaluating proposals are listed below.

- ***Best Value.*** In the best-value approach, a set of evaluation criteria (typically four to five categories) is developed. Each category is weighted to signify both its importance and how much it counts in the evaluation. Although agencies make their best attempts at systematizing this process, this can be a very subjective approach and could lead to an arduous process of contentious evaluations. This process should be considered if the project has specific high-priority issues, such as tight schedules, historic building issues, environmental issues, or specific technical issues.
- ***Low Price, Technically Acceptable (LPTA).*** In the low-price, technically acceptable approach, the proposals first are reviewed for technical acceptability and the price information is excluded. Proposals that meet or exceed the technical acceptability requirements continue in the process to then compete on price alone.
- ***Low Price.*** In the low-price approach, the developer that offers the lowest cost proposal (e.g., installed cost; projected annual production in kWh) and meets minimum requirements (e.g., equipment-quality specifications) is chosen.

Due Diligence

Describing how to handle due diligence on the developer's part has important contractual implications. Due diligence is the effort that a developer must put forth to fully understand the project and the risk of any unknowns that could arise. Contractually, the developer desires recourse if something unexpected comes up that is outside of its ability to perform or that will cause significant cost increases (such as roof structural issues, or discovery of subterranean rock that precludes using standard trenching machines). In such circumstances, the developer might want to be able to walk away from the project or have the option to renegotiate. Agency options range from giving developers what they want contractually to telling developers to factor the risk into their proposals and handle any unexpected circumstances that arise.

Depending on the project, due diligence can require considerable effort and expense. If this is the case with the project, or if this becomes apparent as the RFP process progresses, consider adding steps to the process. Additional steps can include an initial proposal review and the creation of a short list of prospective developers. Those on the short list are invited to continue with the RFP process, which includes expending additional effort and incurring more expense for due diligence. The purpose of this extra step is to assure the developers on the short list that they have a good chance at being successful, and that it is worthwhile to put forth the extra effort and expense required for due diligence. A site due diligence date—after which the developer will have entered into an irrevocable contract—should be specified.

Proposal Process Administration

The description of proposal administration includes the timing and location of proposal meetings and site visits, and the process for answering questions. The proposal meeting is the forum for presenting the project requirements in detail to interested developers and for developers to ask questions. The site visit enables interested developers to assess site conditions and to ask additional questions. The site visit can be held in conjunction with the proposal meeting or be conducted separately. Depending on site conditions and the process chosen, additional site visits could be necessary for respondents to perform additional due diligence.

Questions that arise during and after the proposal meeting and site visit must be handled such that all developers have access to the same information. There are many variations on how this can be accomplished. A recommended method is to write down every question, answer each, and post the questions and answers on a Web site. This process can help avoid variation in answers, minimize participant misunderstanding, and eliminate the possibility of one party receiving more information than another.

Post-Commissioning Performance Specification

Consider adding a post-commissioning performance guarantee into the RFP. This guarantee ensures a minimum level of performance for a specified time after the renewable energy system has been commissioned. An example of this is a PV system that requires quarterly performance verification for the first year of service and which has a guarantee that the system output is to be at least 80% of calculated output based on actual solar insolation (solar energy) for the period. Sample performance specifications and SOWs are available at https://www.fbo.gov/download/ d22/d22e9 1 f03f8ff7af7 1 05083a2b0b7f94/ ARRA_Roof_Final_Specs_9-28-2009.pdf (accessed June 9, 2010); and https://www .fbo.gov/files/archive/ade/ ade3 1d57bafa783a18f8d802437a060a.pdf (accessed July 19, 2010).

4. Issue a Request for Proposal

After the RFP is complete, announce it somewhere that developers can find it. One prominent Web site that is familiar to most developers is Federal Business Opportunities, www.fbo.gov (accessed June 8, 2010). Another Web site option is the Green Power Network, http://apps3.eere.energy financial/ (accessed June 8, 2010). If using the GSA schedule, then it typically is desirable to receive proposals from three different vendors on the schedule. See www.gsaelibrary.gsa.gov/ ElibMain/sinDetails.do?executeQuery=YES&scheduleNumber=56&flag=&filter=&special ItemNumber=206+3 (accessed June 8, 2010).

5. Administer the Request for Proposal

After the RFP is issued, follow the defined timeline and described RFP process. Adjust both as necessary if unforeseen events arise. This step may include site visits, pre-bid meetings, and correspondence related to the project's questions and answers.

6. Evaluate the Proposals

Assemble a small team to evaluate the proposals received. The number of team members to include depends upon the specific project, but the team should have at least three people. Most of the people on this team probably will come from the project team. Other key people to consider including on this team are the:

- Energy manager,
- Facilities manager,
- Project manager,
- Site manager (if there are managers for different areas of a site),
- Solar expert, and
- Utility representative (if a good relationship with the utility exists, then a representative could be included as a non-voting member).

The process for evaluating the proposals should have been established as the RFP was being developed. It is recommended that the merit-review sessions be set up well in advance, to ensure the availability of key personnel. Follow the proposal evaluation criteria described in the RFP and, from the start, clearly define the meaning of each criteria and score. Each agency is also likely has its own review process to follow, and it is important to address that as well.

Evaluating assumptions and exclusions included in the proposal requires particular care. Are the assumptions and exclusions reasonable, based on the information available about the project? Do they demonstrate good judgment? What should be avoided are the costly change orders or price increases that can come with the low cost proposals that are based on poor assumptions or excessive exclusions. The risk is that what initially seems to be the lowest cost proposal actually could be much more costly in the end.

One option to help ensure system performance is to include performance verification as part of the contract. Performance verification should extend for a specified period after commissioning, and the verified performance should meet a predetermined threshold.

7. Award the Contract

There are several options that can be used in awarding the contract.

- **Award Based on Proposal.** The contract is awarded solely on the merits of the proposal as they are determined when the proposals are evaluated.
- **Award with Discussions.** The contract is awarded on the merits of the proposal but is contingent, in part, on further discussions to clarify understandings, agreements, and responsibilities.
- **Award with Discussions and Negotiation.** The contract is awarded on the merits of the proposal but is contingent on further discussions and negotiations. This can be used in the case of receipt of a good proposal that requires adjustments to meet the specific needs of the project. This approach can be employed when unanticipated ambiguities in the RFP or project specifics arise during the RFP process and result in varied proposals that do not quite meet the objectives.
- **Award with Best Proposal.** In this process, a short list of developers is created based on their proposals. The short-list contractors then are asked to develop their best and final offer (BAFO). This request for a BAFO can include information such as updated pricing and design specifications. At this point it is assumed that developers on the short list are technically competent, therefore the evaluation is based primarily on price unless the best value method is employed. This approach also can be used in conjunction with projects that have significant due diligence issues, as noted in the "Develop a Request for Proposal" section. It is recommended that the short list include not more than three developers. As the number of developers on the short list increases, the odds of success decrease for each individual developer as does the willingness of the developers to expend money and effort to tighten their proposals.

8. Design the Project

After the contract is awarded the project design phase begins. The design parameters that the system designer will work within should be clear from the RFP, any questions that arose during the RFP process, and the due diligence performed by the developer. The design kickoff should confirm these design parameters for all parties. It is recommended that design reviews be performed by a third-party, qualified solar-design expert at 25%, 50%, and 100% design completion, to confirm that site requirements are met. When utility interconnection agreements are part of the project, it is recommended that the utility also reviews and approves the project design.

If the design-bid-build contract type is employed, then the system design already has been completed. The developer, however, could have recommendations on design changes that would improve the system. Any changes should be reviewed by a third-party, qualified solar- energy system expert to help confirm that no unanticipated consequences will occur due to the change. A thorough design review always is faster and less expensive than fixing design flaws later. To help with the design reviews, a "Photovoltaic Project Design Evaluation Checklist" is included in Appendix B. The SRCC Web site has links to extensive design guidance for commercial-scale solar heating and cooling systems at www.solar-rating.org/commercial/ guidelines.htm (accessed June 8, 2010).

USPS Marina Mail Processing Center, Los Angeles, California. (Credit: Bill Golove, PIX 11059)

9. Construct the Project

The actual construction of the project typically is not much different from a standard mechanical electrical construction project. If the project includes PVs, then more utility involvement might be required because of the electrical-generation component. This should not be a problem if the utility has been involved with the project from its early stages and has approved the plan.

10. Commission the System

When the system is significantly complete and operational, it is recommended that it be commissioned and inspected by a third-party expert. This is where superior system design and solid performance specifications help to ensure good value. A top-notch commissioning makes certain that the system has been installed properly and is operating to specifications. It also confirms that there are no apparent safety issues due to poor installation (e.g., damaged wire insulation, unprotected high-voltage connections). It also is recommended that a good commissioning plan be established and agreed upon during the RFP process. The commissioning plan can be written into the RFP or be proposed by the developer. If proposed by the developer, then the agency solar expert should review it and make sure that it meets all requirements. A "Photovoltaic Commissioning Checklist" is located in Appendix B.

11. Post-Commissioning Performance

If the agency owns its system, then it is concerned with how it will operate over time and what recourse is available if the system ceases to operate according to expectations. Equipment manufacturers warranty their products and developers might provide a warranty on the system. The key questions are, "What does the warranty cover?" and "How can it be determined whether there is a problem if there is no obvious malfunction?" It is recommended that there be clear agreement with the developer regarding system performance expectations and what constitutes a system failure.

One option to help ensure system performance is to include performance verification as part of the contract. Performance verification should extend for a specified period after commissioning, and the verified performance should meet a predetermined threshold. An example of this is a PV system that requires quarterly performance verification for the first year of service, and a contractual mandate that system output must be at least 80% of calculated output based on actual solar insolation for the period. If desired, the project team and solar expert can develop a reasonable agreement with the developer for a guarantee of this nature.

Long-term monitoring of the system to understand reliability and operations and maintenance costs also is an important part of continued performance and economic benefits. The DOE SETP can track performance and reliability of system installations. If interested in having the DOE track a specific system, contact Michael Quintana at Sandia National Laboratories (maquint@sandia.gov).

POWER PURCHASE AGREEMENT PROJECT

Power Purchase Agreements (PPAs) have been used to finance solar projects since 2003 and they are now driving most commercial solar installations. They are increasingly being utilized by the federal sector. Under a PPA, a private entity (typically a group consisting of developers, construction companies, and finance companies) installs, owns, operates, and maintains customer-sited (behind the meter) solar energy generation equipment. The site purchases electricity or thermal energy through a long-term contract with specified energy prices. Payment is based on actual energy (kilowatt-hours or therms) generated from the solar equipment and consumed by the site. So far, PPAs only have been applied to electricity purchases, but there is no obvious reason why they couldn't be used to purchase thermal energy as well. Be aware that some of the obstacles to PPAs—such as their legality in certain states—does not apply to thermal projects because thermal energy is not regulated in the same manner as electricity production.

A PPA is a relatively new contracting option and, as such, the PPA section of this guide is based on a limited level of federal PPA experience. This section will be updated periodically to reflect new information and recommended best practices—especially if long-term renewable contract authority legislation is passed to make PPAs more financially viable within the federal sector (PPAs almost always require long-term contracts to make the offered price of energy competitive). To address the contract length limitation, federal agencies are exploring methods that are available under existing federal laws and regulations, and also are making other contractual issue improvements. Agencies and industry are encouraged to work to find successful solutions and to share any lessons learned.

Note that innovative options are being used to reduce the transaction costs of completing PPAs, such as multi-award contracts (MACs) that are indefinite delivery, indefinite quality (IDIQ) contracts made with preapproved solar developers. In this example, only a smaller set of project specific details must be worked out and several of the steps listed above can be shortened or can be skipped altogether. An example of this approach is the MAC that the

Naval Facilities Engineering Command Southwest (NAVFAC SW) division is pursuing with five solar developers for PPA projects in its region, and which only can be used by U.S. Navy and Marine Corps facilities.

Steps to Follow

1. Address power purchase agreement-specific issues
2. Select a contracting agent (if needed)
3. Develop and issue a request for information (optional)
4. Develop and issue a request for qualifications (optional)
5. Develop a request for proposal
6. Issue a request for proposal
7. Administer the request for proposal
8. Evaluate the proposals
9. Award the contract (issue any needed indefinite delivery, indefinite quantity [IDIQ] task order)
10. Design the project
11. Construct the project
12. Commission the system
13. Monitor the performance period
14. End contract oversight

PROS	CONS
• Renewable energy developer is eligible for tax incentives and accelerated depreciation, which could lead to reduced energy costs. • Agency is not required to provide up-front capital. • Renewable energy developer provides operations and maintenance for the duration of the contract (no agency O&M responsibilities). • Government faces minimal risk. • Agency typically receives a known long-term electricity or thermal energy price for a portion of the site load (which reduces the price risk of fluctuating utility energy prices). • Developer has incentive to maximize production by the system (compared to the case of a direct purchase of the system). • Agency potentially can use available funds for a front-end buy down to get a better PPA price or a larger system.	• Transaction costs include a significant learning curve and time investment. • Federal-sector experience is limited. • Civilian agencies are limited to 10-year term PPA utility contracts (the U.S. Department of Defense [DOD] has 2922A authority, which permits 30-year terms). • Site-access issues are complex. • Management and ownership structures are complex. • Contract termination penalties.

> **Case Studies**
>
> (See Appendix C)
>
> - NREL PV Project, Golden, Colorado
>
>   720-kW PV
>   www.eere.energy.gov/femp/pdfs/ pfs_mesatoparray.pdf
>
> - Fort Carson PV Project, Colorado Springs, Colorado
>   2-MW PV
>   www.3phases.com/news/news-item. php?id=32

1. Address Power Purchase Agreement–Specific Issues

Before beginning the power purchase agreement process, confirm that a PPA is allowed in the state in which the project is located. The restriction information for PPAs is available on the DSIRE Web site at www.dsireusa.org/summarymaps/index.cfm?ee=1&RE=1 (accessed July 19, 2010). If DSIRE indicates that PPA status is unclear or apparently disallowed, it is recommended that the state's energy office or public utility commission be contacted to help determine whether a PPA is legal for the site.

In general, PPAs typically are used only to implement larger projects (typically 100 kW or greater). This is based on several cost factors, including transaction costs, securing financing, and economies of scale that make the PPA electric price more acceptable. Recently, however, there have been indications that developers might consider smaller projects. For a relatively small project, several options exist. Multiple smaller solar projects can be aggregated into one larger project, could be bundled with energy efficiency in an ESPC or UESC, or can use agency funding. Generally PPAs can be used for a solar thermal project, but the assumption is that most solar thermal projects are not large enough to interest developers.

To be economical, most PPAs require long-term contracts—generally 15 to 20 years—and most agencies do not have the authority to enter into utility contracts of this length. Congress might change this but at present a workaround is required. Western Area Power Administration (Western) can help with long-term contracts for sites in its area (www.wapa.gov; accessed June 8, 2010). Western can negotiate and sign the PPA on behalf of a federal agency, but the federal agency actually must select the solar developer.

Innovative methods to address the contract length limitations are being explored. An example is a long-term land-use agreement that includes a provision requiring the solar project developer to give the federal agency hosting the solar project right of first refusal on purchase of the power at a predetermined price. If the agency does not purchase the power, then the developer is free to sell it to the local utility. Before utilizing this method, investigate legal issues and determine any possible effect on a developer's proposed electricity price. The developer could perceive more income risk and increase the price of electricity to compensate.

The secretary of the DOD has the authority (10 U.S.C. §2922A) to allow long-term contracts of up to 30 years in duration (with approval of the secretary of defense). The U.S. Navy plans to use this authority for the NAVFAC SW division multiple-award contract. Approval through the U.S. Secretary of Defense, delegated to the U.S. Secretary of the Navy, will be sought under task order awards for these projects.

2. Select a Contracting Agent

Determine the best contracting route to use. Typical options include local, regional, or headquarters contracting staff, Defense Logistics Agency (DLA Energy, formerly Defense Energy Support Center), or Western Area Power Administration. Contact the contracting agent early after project identification to determine the best approach for the next steps in the process. Note that Western only signs the PPA, the agency for the site must select the solar developer.

3. Develop and Issue a Request for Information (*Optional*)

A request for information (RFI) is a way to obtain feedback on the proposed project to help refine and develop the RFP. Recommendations of types of projects for a specific site typically are helpful. The U.S. Navy used an information-request process in California, and respondents overwhelmingly recommended behind-the-meter PV projects sized to meet on-site loads. This is due to local interconnection requirements for large systems connecting directly into the grid. The information was used to refine the government's requirements for the scope of work used in the RFP. An RFI also allows industry to comment on the proposed process.

4. Develop and Issue a Request for Qualifications (*Optional*)

Another optional step that has been used for at least one federal site is a request for qualifications (RFQ). The purpose of the RFQ is to obtain a list of developers that are interested in the project and to learn about their specific qualifications. Developers that meet a stated qualification level can submit a proposal based on the RFP created in the subsequent step in the process. Developers typically decide which RFPs to respond to based on the limited development funds available. Developers have indicated a preference for this step because responding to an RFQ is relatively easy and inexpensive, and it reduces the field of competitors. A smaller field of competitors increases the probability of success, and qualified developers are assured that they are competing against other qualified developers. Receiving proposals only from qualified developers also can reduce the team's review workload and encourage qualified developers to invest more in their proposals, as there is a greater chance of being awarded the contract. The criteria to be used to qualify proposers must be stated. If the RFQ step in the process is not used, then the information that would have been received in the RFQ must be requested in the next step, the RFP. The following list includes items to consider including in the RFQ.

- Executive bios
- Letters from investors
- Professional affiliations
- Project experience (e.g., size, type, year built, customer)
- References
- Sample PPA
- Three years of audited financial statements
- Evaluation criteria and/or evaluation process, if selecting a short list of proposers

5. Develop a Request for Proposal

A request for proposal is the document issued to the public to solicit proposals; in this case, from solar developers. The RFP describes how the proposal process is to be conducted and provides information that can be used as a basis for a developer's proposal. Sample documents can be found at the FEMP PPA Web site (www.eere.energy financing/ppa_sampledocs.html; accessed July 19, 2010). An RFP should include the following elements (listed alphabetically, and not in order of importance).

- Assignment of renewable energy attributes (ownership of the RECs)
- Drawings and maps (if available)
- Building restrictions (for rooftop installations) such as roof penetrations, warranty considerations, roof replacement plans, and maximum roof weight capacity
- Contracting officer representative information (if applicable)
- Current energy-consumption data
- Infrastructure requirements (if any), such as roads, fences, electrical system upgrades, tree removal, and determining which party is responsible for coordination and payment
- Environmental requirements such as NEPA, NHPA, Endangered Species Act, and other applicable federal, state, and local requirements
- Due diligence
- End-of-project options
- Evaluation criteria and process
- Land-use agreement (include this as an attachment)
- Language of the PPA (optional)
- Limits on proposed project timeline
- Liquidated damages
- Proposal process administration plan (e.g., proposal meetings, site visits, answers to questions)
- Qualifications (if RFQ step is not used)
- Required submittals (can include acceptable pricing formats; for example, fixed-price only with escalation)

- Restrictions (or preferences) on parties allowed to submit proposals (e.g., small business, woman owned, veteran owned); this is a policy decision that can be based on agency- or site-level preferences
- Safety restrictions for construction
- Site addresses
- Site design criteria
- Site fire standards and safety requirements
- Specific site-access requirements
- Solar energy system requirements, including system type and location
- Termination for convenience (provisions and termination schedule)
- Timelines for proposal process

Assignment of Renewable Energy Attributes

The RFP must be clear on ownership of the attributes of the renewable energy generated. Depending on the markets for these attributes, they can be a major factor in determining the PPA electricity price. These attributes also might factor into the agency's renewable energy goal requirement (for information, contact the person responsible for such requirements). It is important to be clear that renewable energy credits include GHG emissions and all other environmental attributes. If the renewable energy credits are sold, then replacement RECs can be purchased for credit towards the EPAct 2005 renewable energy goal. Present guidance is that no credit towards the E.O. 13514 goal is allowed if the RECs (including the GHG emissions) are sold or are retained by the developer.

Criteria and Process for Evaluating Proposals

Describing the criteria and the process to be used to evaluate the proposals helps developers to structure their responses and ensures that the project's primary issues are addressed. This also makes it easier to review proposals. The respondents address identified criteria in separate sections, eliminating the need for a reviewer to pick statements from the entire proposal that apply to the criteria. The three common processes used for evaluating proposals are listed below.

- *Best Value.* In the best-value approach, a set of evaluation criteria (typically four to five categories) is developed. Each category is weighted to signify both its importance and how much it counts in the evaluation. Although agencies make their best attempts at systematizing this process, this can be a very subjective approach and could lead to an arduous process of contentious evaluations. This process should be considered if the project has specific high-priority issues, such as tight schedules, historic building issues, environmental issues, or specific technical issues.
  - Low Price, *Technically Acceptable.* In the LPTA approach, the proposals first are reviewed for technical acceptability and the price information is excluded. Proposals that meet or exceed the technical acceptability requirements continue in the process to then compete on price alone.
- *Low Price.* In the low-price approach, the developer that offers the lowest cost proposal (e.g., installed cost; projected annual production in kilowatt-hours) and meets minimum requirements (e.g., equipment- quality specifications) is chosen.

Under a PPA, a private entity (typically a group consisting of developers, construction companies, and finance companies) installs, owns, operates, and maintains customer-sited solar energy generation equipment.

POWER PURCHASE AGREEMENT PROJECT CASE STUDY

NREL Mesa Top PV (DOE)

In order to meet its EPAct 2005 goal of 7.5% of electricity from renewable sources, NREL, in Golden, Colorado, utilized a power purchase agreement to build a PV system on top of the mesa above its campus. The system is a 720 kW, ground mounted, single axis tracking array that produces 1,200 MWh annually or more than 7% of the lab's electricity use in 2008. NREL selected SunEdison to do the project. Western Area Power Administration used its long-term contract authority to sign a 20-year solar power and services agreement with SunEdison to provide solar-generated electricity to DOE for use at NREL. SunEdison financed, built, owns, and operates the PV system. The company sells the RECs to Xcel Energy, and NREL purchases replacement RECs to maintain its carbon-neutral status and meet federally-mandated renewable energy goals. The price per kilowatt-hour for electricity from the array is equal to or less than the price NREL pays for utility-supplied power. There were no up-front costs to NREL, and the EPAct goal was achieved four years ahead of requirement. SunEdison provides data on system performance that help NREL track target achievement. The system has been operational since December 2008.

Different variations on these three evaluation processes have been created in an effort to elicit a better proposal and to reduce the investment required in the proposal process. One example essentially is an RFQ combined with an indicative price. The RFQ addresses the technical acceptability of the developer. The indicative price is meant to provide a "reasonable" idea of the price terms, and requires less developer investment than that required for a full proposal. Based on the evaluation of the proposals, one or two respondents are chosen to submit a full proposal.

Evaluation criteria can include the following.

- Amount of energy generated on an average hourly, monthly, and annual basis over the term of the agreement, including a degradation factor
- Developer's experience and performance track record and references
- Developer's financial health
- Developer's ongoing long-term ability to service the system
- How specific site issues and requirements will be addressed
- Implementation plan
- Local sourcing of components and labor
- Price
- Performance plan
- Quality components
- Quality plan

Due Diligence

Describing how to handle due diligence on the developer's part has important contractual implications. Due diligence is the effort that a developer must put forth to fully understand the project and the risk of any unknowns that could arise. Contractually, the developer desires recourse if something unexpected comes up that is outside of its ability to perform or that will cause significant cost increases (such as roof structural issues, or discovery of subterranean rock that precludes using standard trenching machines). In such circumstances, the developer might want to be able to walk away from the project or have the option to renegotiate. Agency options include—but are not limited to—giving developers what they want contractually or telling developers to factor the risk into their proposals and handle any unexpected circumstances that arise. Government agencies do have language for equitable adjustments in price given increased scope of work, however. Also, in accordance with Federal Acquisitions Regulation (FAR) 15.208(e), a contractor has the right to withdraw its proposal at any time prior to the contract award. After award, this becomes a contract-termination issue.

Land-Use Agreements

Land-use agreements govern the site access given the developer during the term of the project. There can be two phases addressed by this agreement, the construction phase and the production phase. The site-access requirements during these two phases can be quite different, which is why they could have different limitations and may be handled separately. The term of the power purchase agreement could start after construction is completed, therefore certain terms and conditions of the PPA might need to be included in the construction phase. Land-use agreements typically take the form of leases, easements, licenses, or land purchase. Leases are for a limited term and usually are for exclusive occupation rights. Easements are a nonexclusive right to occupy and cross a property and primarily are irrevocable. Licenses typically are easier to use and have been used by several agencies for rooftop systems. Another option is for the solar developer to purchase nearby nonfederal land as the location for the system. When considering land-use agreement options, work with the team's legal counsel and real property staff, because agencies have different requirements. Sample land use agreements are available at: www.eere.energy sampledocs.html (accessed July 19, 2010).

Language of a Power Purchase Agreement

It is recommended that preferred PPA language or key legal considerations for the contract agreement be included in the RFP, as this affects the proposals submitted by developers. The power purchase agreement language incorporated in the RFP can be negotiated. It should, however, cover the pricing request for the term of the agreement (with escalators), termination fees for each year, and disposition of equipment at the end of the contract term. Experience has shown that RFPs that do not contain this language have produced administrative problems due to ambiguity.

System purchase options can be included if the agency foresees that it might want to purchase the system before the end of the PPA. This usually is considered after the sixth year of the project, after all the investment tax credits and accelerated depreciation benefits are exhausted. Buyout provisions are based on fair market value (FMV) or the present value of income expected from the remaining life of the PPA. The advantage of owning the system is the owner does not have to pay for the power produced. If an owner has not signed an O&M contract, however, then that owner is responsible for the operation and maintenance costs. At the end of the PPA contract, the agency could purchase the system for fair market value, extend the PPA (if allowed), issue a follow-on RFP, have the contractor abandon the system in place, or have the contractor remove the system.

Liquidated Damages

It is recommended that the contract include a clause regarding liquidated damages (LD). This clause should address and develop a method for calculating damage payments for the failure to perform contractual obligations. If failure to meet obligations has clear and quantifiable monetary consequences, then damages can be calculated easily. Damages that arise from failures that do not have clear and quantifiable monetary consequences should be addressed by this clause. Most examples of situations to which an LD clause would apply are related to not meeting stage-gate requirements. Stage-gate requirements might include:

- Design submissions,
- Permitting,
- Procurement of modules, and
- Commissioning or completion.

Proposal Process Administration

The proposal administration description includes the timing and location of proposal meetings and site visits, and the process for answering questions. The proposal meeting is the forum for presenting the project requirements in detail to interested developers and for developers to ask questions. The site visit enables interested developers to assess site conditions and to ask additional questions. The site visit can be held in conjunction with the proposal meeting or be conducted separately.

Depending on the site conditions and the process chosen, additional site visits could be necessary for respondents to perform additional due diligence.

Questions that arise during and after the proposal meeting and site visit must be handled such that all developers have access to the same information. There are many variations on how this can be accomplished. A recommended method is to write down every question, answer each, and post the questions and answers on a Web site. This process can help avoid variation in answers, minimize participant misunderstanding, and eliminate the possibility of one party receiving more information than another.

Required Design and Construction Submittals

The RFP should be clear on what submittals are required from the developers for the proposals and, if successful, what is required during the design and construction of the system. The proposal submittal could include:

- Conceptual layout;
- Detailed project implementation plan;
- Line diagram (electrical schematic diagram);
- Projected energy performance (average hourly, monthly, and total with degradation factored over the term of the agreement); and
- System components, such as modules, inverters, and racking, with specifications and warranty information.

Submittals required during the project construction should include as-built drawings and final system specifications.

System Requirements

The system requirements section of the RFP should include a description of expectations regarding the project and any pertinent information that will help the developers give a solid proposal. Include one or more location options and, if desired, a minimum system capacity for each location. Note whether each PV system will be evaluated separately or as part of a group. System requirements (e.g., PV panel types, configuration) should not be prescriptive and instead should give developers the opportunity to propose a system that is most economical based on individual experience.

Fort Carson PV Project, Colorado Springs, Colorado, 2-MW PV (*Credit: U.S. Army Fort Carson. PIX 17394.*)

The system description should include expected technology type, size or performance range, location, and any site-specific considerations or limitations (e.g., roof penetration preference, roof-replacement schedule). Site information that should be provided, if available, includes pertinent electrical information and drawings, site characteristics (e.g., soil studies, roof plans), site load information (maximum/ minimum demand for each month), consumption information (hourly if available), environmental factors, historic information, interconnection options, acceptable inverter locations, and any other pertinent information. Sample RFPs are available at www.eere. energy. html (accessed July 19, 2010).

6. Issue a Request for Proposal

After the RFP is complete, announce it somewhere that developers can find it. One prominent Web site that is familiar to most developers is Federal Business Opportunities, www.fbo.gov (accessed June 8, 2010). Another Web site option is the Green Power Network, http://apps3.eere.energy financial/ (accessed June 8, 2010), or www.fedconnect.net (accessed June 8, 2010).

7. Administer the Request for Proposal

After the RFP is issued, follow the defined timeline and described RFP process. Adjust both as necessary if unforeseen events arise. This step may include site visits, pre-bid meetings, and correspondence related to questions and answers related to the project.

What should be avoided are costly change orders or price increases that can come with the low cost proposals that are based on poor assumptions or excessive exclusions. The risk is that what initially seems to be the lowest cost proposal actually could be more costly in the end.

8. Evaluate the Proposals

Assemble a small team to evaluate the proposals received. The number of team members to include depends upon the specific project, but the team should have at least three people. Most of the people on this team probably will come from the project team. Other key people to consider including on this team are:

- Energy manager,
- Facilities manager,
- Legal/procurement expert,
- Project manager,
- Site manager (if managers for different areas of site), and
- Solar expert.

The process for evaluating the proposals should have been established as the RFP was being developed. It is recommended that the merit- review sessions be set up well in advance to ensure the availability of key personnel. Follow the proposal evaluation criteria described in the RFP and, from the start, clearly define the meaning of each criteria and score. Each agency also likely has its own review process to follow, and it is important to address that as well.

Evaluating the assumptions and exclusions included in the proposal requires particular care. Are the assumptions and exclusions reasonable, based on the information available about the project? Do they demonstrate good judgment? What should be avoided are costly change orders or price increases that can come with the low cost proposals that are based on poor assumptions or excessive exclusions. The risk is that what initially seems to be the lowest cost proposal actually could be more costly in the end.

When evaluating proposals for pricing options, be aware that if an acceptable pricing structure is not specified in the RFP then many different options could be given. Common pricing structures include escalation factor (usually 1% to 3%), firm- fixed price, utility-rate linked, or a de-escalation factor. An escalator is the percentage that the PPA price per kilowatt-hour will increase annually. A first year price with escalator usually is less than a fixed price but will increase to more than the fixed price during the term of the PPA.

Typically, an evaluation of these pricing structures can be based on lowest present cost for the expected production and term of the project.

The winning proposal should be compared to current utility rates and the expected future rates, based on inflation and discount rates taken from the National Institute of Standards and Technology (NIST)/Energy Information Administration (EIA) "Energy Price Indices and Discount Factors for Life-Cycle Cost Analysis— May 2010, Annual Supplement to Handbook 135" (available at www.eere.energy pdfs/ashb10.pdf; accessed July 19, 2010). Other price forecasts also can be used for comparison purposes to help the site decide whether a contract award is recommended. Other time- and project-specific requirements, such as when funds will be available to pay for production (e.g., a large amount of funds might be available in the first year of the project, but perhaps more significant funds will not be available until a subsequent year) also can be considered. As noted, standby charges and possible utility tariff changes should be compared to historic and projected utility costs.

9. Award the Contract

The process for contract award depends on the type of evaluation used, and could include negotiations with the proposers. Several options can be used for awarding the contract. (Sample contracts are available at www.eere.energy.gov/ femp/financing/ppa_ sample docs.html; accessed July 19, 2010.)

- *Award Based on Proposal.* The contract is awarded solely on the merits of the proposal as determined when the proposals were evaluated. This scenario is very unlikely, as discussion almost always is required.
- *Award with Discussions.* The contract is awarded on the merits of the proposal but is contingent in part on further discussions to clarify understandings, agreements, or responsibilities.
- *Award with Discussions and Negotiation.* The contract is awarded on the merits of the proposal but is contingent on further discussions and negotiations. This can be used in the case of receipt of a good proposal that requires adjustments to meet the specific needs of the project. This approach can be employed when unanticipated ambiguities in the RFP or project specifics arise during the RFP process and result in varied proposals that do not quite meet the objectives.
- *Award with Best Proposal.* In this process, a short list of developers is created based on the proposal. Those on the short list then are asked to develop their best final proposal revision (FPR). This request for FPR can include information such as updated pricing and design specifications. At this point it is assumed that developers on the short list are technically competent, therefore the evaluation primarily is based on price unless the best value method is employed. This approach also can be used in conjunction with projects that have significant due diligence issues, as noted in the "Develop a Request for Proposal" section. It is recommended that the short list include not more than three developers. As the number of developers on the short list increases, the odds of success decrease for each individual developer, as does the willingness of the developers to expend money and effort to tighten their proposals.

10. Design the Project

After the contract is awarded the project design phase begins. The design parameters that the system designer will work within should be clear from the RFP, the final revised proposal, any negotiations that occurred during the RFP process, and due diligence performed by developer. The design kickoff should confirm these design parameters for all parties. It is recommended that design reviews be conducted by a third-party, qualified solar-design expert at 25%, 50%, and 100% design completion stages, to confirm that requirements are met. A thorough design review always is faster and less expensive than fixing design flaws later. To help with the design reviews, a "Photovoltaic Project Design Evaluation Checklist" is included in Appendix B.

11. Construct the Project

During project construction, the primary considerations are regarding coordinating with the developer. To enable a successful coordination during this phase, first identify a single point of contact. Major areas of coordination include the timing of work (particularly if construction could interfere with the site's mission), ensuring that critical deadlines are met (especially those regarding incentives), assisting with interconnection issues (including interconnection and net metering agreements), and handling incentive applications. The final piece of the construction process is the commissioning of the system—which makes possible the system interconnection and start up.

A good commissioning ensures that the system has been installed and is operating to specifications. It also confirms that there are no apparent safety issues due to poor installation (e.g., damaged wire insulation, unprotected high- voltage connections).

12. Commission the System

Although the system is owned and operated by a third party and the agency is just purchasing the energy output of the system, the system still is located on the agency's site and the agency has an interest in how well the system performs (for the credit toward renewable energy targets). Commissioning of the system is the responsibility of the solar developer; however, it is recommended that the agency be aware of any issues and reports resulting from the commissioning. A good commissioning ensures that the system has been installed and is operating to specifications. It also confirms that there are no apparent safety issues due to poor installation (e.g., damaged wire insulation, unprotected high-voltage connections). There could also be requirements from the REC purchaser that must be met. A "Photovoltaic Commissioning Checklist" is provided in Appendix B.

13. Monitor the Performance Period

The primary duty during the PPA's performance period is to track actual production and pay for electrical production. Operation and maintenance usually is the responsibility of the developer. If the developer owns the RECs produced by the project and the agency needs to meet renewable energy targets, then it can purchase replacement RECs every year.

If the developer agrees to long-term monitoring of the system, then the DOE Solar Energy Technologies Program can track performance and reliability of system installations. For information on having the DOE track a specific system, contact Michael Quintana at Sandia National Laboratories (maquint@sandia.gov).

14. End Contract Oversight

The end of the power purchase agreement is characterized by the decision on the preferred system purchase or other termination option (discussed above in "Language of a Power Purchase Agreement" section). This could be oversight of the system removal, extension of the PPA, or purchase of the system and continuation of the O&M. The choice of options most likely will be influenced by circumstances existing 20 years in the future, and this should be taken into account when developing the options at the start of the project.

ENERGY SAVINGS PERFORMANCE CONTRACT

Energy savings performance contracts (ESPC) have a long history of use in the federal sector and have primarily been used for energy efficiency projects. They are increasingly being seen, however, as a long-term financing method for solar projects. An ESPC is a guaranteed savings contracting mechanism that requires no up-front cost. An energy services company (ESCO) incurs the cost of implementing a range of energy conservation measures (ECMs)—which can include solar—and is paid from the energy, water, and operations savings resulting from these ECMs. The ESCO and the agency negotiate to decide who maintains the ECMs. Payments to the contractor cannot exceed savings in any one year. These contracts are recommended for renewable energy projects only if energy-efficiency measures also are being performed.

Multiple contracting options are available to agencies interested in an ESPC. The DOE offers an indefinite delivery, indefinite quantity contract designed to make an ESPC as cost-effective and easy to implement as is possible for federal agencies. Several ESCOs are prequalified and have accepted the terms of the IDIQ contract; these companies thus can respond to project requests. The U.S. Army also has an IDIQ contract in place as an alternative to the DOE option. The discussion below focuses on the DOE ESPC process, which is explicitly defined. Additional information is available on the FEMP Web site at www.eere.energy (accessed June 10, 2010). Also, the FEMP has extensive resources including contract templates, flowcharts, and process guidance on its Web site at www.eere.energy.gov/femp/financing/espcs_resources.html (accessed July 19, 2010).

Steps to Follow

1. Plan the project
2. Perform a preliminary assessment and ESCO selection
3. Perform investment grade audit to award
4. Design the project
5. Construct and install the system
6. Commission the system
7. Monitor the performance period
8. Perform project close out

Case Studies

(See Appendix C)

- **U.S. Marine Corps Base PV Project, Twenty-nine Palms**
  www.bp.com/liveassets/bp_internet/solar/bp_solar_usa/STAGING/local_assets/downloads_pdfs/29PalmsTag.pdf
- **Federal Correctional Institution, Phoenix, Arizona**
  Parabolic trough solar water-heating system www.eere.energy

---

These contracts are recommended for renewable energy projects only if energy-efficiency measures also are being performed.

PROS	CONS
• The 25-year contract length fits well with longer renewable energy paybacks. • The performance is guaranteed. • The operations and maintenance can be included as part of the contract. • The agency in charge of the site can require that solar be a part of the project. • A project facilitator is assigned (FEMP funded through initial proposal or preliminary assessment). • The sale of excess electricity and thermal energy is allowed (EISA provision). • The agency contracting officer (CO) has the discretion to allow ESCO or third-party ownership of the renewable energy conservation measures eligible for federal and state tax incentives.	• Since ESCOs traditionally do not own assets, it is difficult to monetize tax incentives related to solar. • Not recommended for renewable-only projects.

1. Plan the Project

The DOE Super ESPC (a DOE IDIQ contract with approved energy services companies) requires the involvement of a federal financing specialist (FFS) and a project facilitator (PF).

The services of the FFS are provided at no cost throughout the project. The services of the PF are provided by the FEMP at no cost up through agency review of the preliminary assessment. When further PF services are required, they are contracted on a reimbursable basis for labor and travel costs. Once the FFS and PF have been identified, assemble the site team, FFS, and PF and put together a notice of opportunity that is to be sent to all energy services companies on the approved list.

The notice of opportunity can be as simple as a one-page letter that gives a summary of what might be included in the project and includes a request for a response from interested ESCOs. The notice of opportunity could include site data for known energy-system improvements, indicate the desire for renewable energy projects, include a schedule of the ESCO site visits, and provide the timeline for submission of a preliminary assessment (PA). It is recommended that the project scope be open to all types of projects. As in any federal procurement, it is important that fair opportunity be given to all potential contractors, especially if large projects materialize after some ESCOs are removed from consideration.

If details are provided in a notice of opportunity or site data package, the results of a renewable energy screening may be included. NREL offers a renewable energy optimization screening at no cost to federal agencies upon request. Other energy efficiency screening reports may also be included. (for more information, see the FEMP ESPC Web site at www.eere.energy.gov/femp/financing/espcs_techplanning.html; accessed June 8, 2010). Site information for other systems that should be targeted for upgrades can be included along with utility usage data. Requested information from the ESCO could include qualifications, past performance, and markups—these can help in the ESCO-selection process that the project team must develop. The ESCOs that are interested in the project submit the requested information to the agency's contracting officer's representative (COR). The team evaluates the responses and one or more ESCOs to proceed to the next step.

2. Perform a Preliminary Assessment and ESCO Selection

The beginning of the preliminary assessment (PA) phase is a kickoff meeting between the site team and the selected ESCOs. A preliminary site assessment follows the kickoff meeting. This assessment identifies the renewable energy and energy-efficiency measures to be considered for the project. The ESCOs develop preliminary assessments (proposals) which the team then reviews. Sample criteria for evaluating a PA can be found online at www.eere.energy.gov/femp/pdfs/4_4_ preliminaryassessmentguidance.pdf (accessed July 19, 2010).

The ESCO selection can occur before or after the preliminary assessment depending on ESCO responses in Step 1(Plan the Project). Contracting officers (CO) are given broad discretion in the DOE IDIQ as to their contractor selection approach. Regardless of the approach and the timing of the selection, fairness must be demonstrated to all contractors and the approach must follow section H.3 of the DOE IDIQ. Aside from some exceptions to "Fair Opportunity", the agency CO must consider price in the selection decision. When the agency downselects to a single ESCO, the agency CO must document the basis for the downselection. The DOE IDIQ contract can be found online at www.eere.energy.gov/femp/pdfs/generic_idiq_ espc_contract.pdf (accessed August 20, 2010)

During this phase, the agreement for continuing the PF services—which are required for the ESPC process—must be finalized. These services are estimated to cost between $50,000 and $75,000 for an average project.

3. Perform an Investment Grade Audit to Award

The IGA is the detailed assessment of prospective energy and water projects. This audit determines the economic viability and bankability (investor financing requirement) of the project. The audit is a joint effort between the ESCO and the agency team, and is characterized by ongoing negotiation. Items that are a product of this effort and must be agreed upon include baseline usages, correct calculation methods, and appropriate measurement and verification (M&V) procedures. The result of this effort is a final proposal that is produced by the ESCO. Note that all ESCO costs up to this point are borne by the ESCO, and that the IGA is a significant effort that can cost $1,000,000 or more (for complex projects). Agency review of the proposal and final negotiations precede the award of the task order.

ENERGY SAVINGS PERFORMANCE CONTRACT CASE STUDY

U.S. Marine Corps Base PV Project, Twenty-nine Palms

The procurement of a 1.1 MW solar energy project was awarded under an Army Corps ESPC agreement to Johnson Controls Inc. in 2002. The military base also received a rebate of $4.5 million from the State of California to buy down the cost of the system. The total cost of the PV project was about $12.6 million after the rebate. As a part of the ESPC agreement, there is an annual guarantee of $400,000 in energy cost savings in conjunction with an estimated 2,500,000 kWh of annual production from the PV system. The PV system was installed as a part of a larger project that included installation of a new cogeneration plant. The project also included upgrades to the energy management system, chilled water plants, and daylighting. Total savings from the overall project to date equal $6.9 million. The PV system utilizes 8,700 modules sited on 6.5 acres. It is tied into the cogeneration plant at 12.47 kilovolt (kV).

Before the contract is completed the task order RFP must be developed. The IDIQ contract language is the default contract language. The task order RFP includes agency-specific contract language, however, that supersedes corresponding IDIQ contract language, and its development can be a significant effort. The final contract consists of three pieces, the task order RFP, the IDIQ, and the final proposal, and it is important to ensure consistency between all three documents. The final result of this phase is the task order award.

Under an ESPC, the title to the improvements installed under the ESPC transfers to the agency upon final project acceptance. For solar energy projects or the solar energy piece of a larger project, the agency contracting officer (CO) has the discretion to allow private-party ownership of renewable energy systems by the ESCO or a third party. With private-party ownership there is an option to implement an energy services agreement as part of the ESPC. To take advantage of tax benefits, an ESA allows a third party (ESCO or investor) to hold title to the solar energy portion of the project and to reap the tax benefits—and enables the third party to pass some of these benefits back to the agency to improve the economics of the project. For projects in which solar energy generation only reduces site load, the energy production is measured and counted toward offsetting utility energy purchases. When entering into an ESA, the agency may also consider buyout or removal provisions that come into effect at the end of the project.

4. Design the Project

After the contract is awarded the project design phase begins. The design parameters that the system designer will work within should be clear from the final proposal. The design kickoff should confirm these design parameters for all parties. It is recommended that design reviews be performed by a third-party, qualified solar- design expert at 25%, 50%, and 100% design completion, to confirm that requirements are met. A thorough design review always is faster and less expensive than fixing design flaws later. To help with the design reviews, a "Photovoltaic Project Design Evaluation Checklist" is included in Appendix B. After the design has been reviewed and accepted, a notice to proceed is issued, and construction can begin.

5. Construct and Install the System

The construction phase of the process is much like that of any other construction project; however, most of the team's work is to coordinate with the construction crews, ensure that the site mission is not unduly impeded, and to meet any milestones and any agency contractual obligations. The contract should be clear on which party is fiscally responsible if obligations are not fulfilled.

If a project is complex, then there could be partial project acceptances that occur as different measures are completed. The motivation for partially accepting projects is to start accruing savings that can be set aside until final project acceptance. Any money set aside due to savings or other agency funding (e.g., funding that was earmarked for upgrades that the ESCO now is performing) can be used to buy down the project financing before final project acceptance. This reduces interest payments over the life of the project and provides potentially significant savings. After the project is accepted, the payments cannot exceed savings and there is no additional option to buy down or prepay the project unless a termination for convenience is exercised. When engaging in partial project acceptance, an agency must commission the measure and implement the M&V protocol to verify the savings before it can start banking any savings.

6. Commission the System

When the solar energy system is significantly complete and operational, the ESCO will commission the system. It is recommended that the agency have a knowledgeable representative present at the commissioning to represent the agency's interest. This is where superior system design helps ensure good value. A top-notch commissioning makes certain that the system has been installed properly and is operating to specifications. It also confirms that there are no apparent safety issues due to poor installation (e.g., damaged wire insulation, unprotected high-voltage connections). The commissioning plan can be written into the final proposal (a "Photovoltaic Commissioning Checklist" is included in Appendix B). It is important to ensure the quality of the installation and to make sure that the measures are operating as expected. When commissioning is satisfactorily complete and post-installation M&V has started, final project acceptance is given and the performance period phase of the project begins.

7. Monitor the Performance Period

The performance period involves measuring ECM performance and verifying savings annually, confirming ESCO or agency maintenance to keep project measures operating as expected, and ensuring agency payments on the contract. The M&V and maintenance activities are carried out as specified in the contract. Any discrepancies or objections to the reported savings must be resolved and then the annual payment can be made. It is important to decide whether to pay in advance or in arrears; the difference equals a year of interest on the project loan.

Phoenix Federal Correctional Institution, Phoenix, Arizona, parabolic trough solar water-heating system. (*Credit: Ed Hancock, Mountain Energy Partnership. PIX 09048.*)

If the developer agrees to long-term monitoring of the system, then the DOE Solar Energy Technologies Program can track performance and reliability of system installations. If interested in having the DOE track a specific system, contact Michael Quintana at Sandia National Laboratories (maquint@sandia.gov).

8. Close Out the Project

At the end of the performance period the task order is closed out. At this point the agency will no longer be required to make payments on the contract.

UTILITY ENERGY SERVICES CONTRACT

Utility energy services contracts (UESC), like ESPCs, have a history of use in the federal sector primarily for energy efficiency projects. Now, these contracts are also being seen as a method of long-term financing, with the added benefit of usually being a sole source contract. A UESC is an agreement that allows a "serving" utility to provide an agency with comprehensive energy- and water-efficiency improvements and demand-reduction services. The utility could partner with an ESCO to provide the installation, but the contract is between the federal agency and the "serving" utility. This contracting mechanism primarily is for bundled energy-efficiency and renewable energy projects, and typically is not used for standalone renewable energy projects. The steps in the UESC process are well defined, but different utilities might describe them differently. The process steps described below are representative of the general process. See www.eere.energy.gov/femp/ financing/uescs.html (accessed June 7, 2010).

An effort currently is underway to define a process for a utility renewable electric service contract (URESC) for parties interested in pursuing a standalone solar electric project with a utility in a specific service territory. The URESC concept is envisioned to produce a cross between a PPA and an UESC. It is hoped that an URESC project will commence in 2010 and define this financing and contracting option.

Note that the following discussion focuses on the renewable energy portion of a UESC project. For general information and assistance with UESCs, the FEMP offers *Utility Energy Service Contract: Enabling Documents*, available at www.eere.energy.gov/femp/pdfs/uesc_enabling_ documents09.pdf (accessed June 7, 2010).

Steps to Follow

1. Introduction: Contract or agreement review
2. Perform a preliminary study
3. Perform an agency review
4. Project implementation proposal (investment grade audit/detailed feasibility study)
5. Negotiate and accept the proposal (construction contract)
6. Detailed design acceptance
7. Construct project
8. Construction acceptance
9. Monitor performance period
10. Close out the project

PROS	CONS
• The UESC contract term is 10 to 25 years, and varies by agency (average project term is 14 years). The EISA (section 513) prohibits agency policies that limit privately financed contract terms to a maximum period of less than 25 years. • The GSA legal opinion states that extended utility agreements are allowed (*Utility Energy Services Contracts: Enabling Documents*, www.eere.energy.gov/femp/ pdfs/uesc_enabling_ documents09.pdf; accessed June 7, 2010). • Utilities now are eligible for a renewable investment tax credit (the utility must own a renewable energy plant). • Interconnection, tariff, and standby issues should be minimal with utility ownership (but this is not always true and should be explored prior to proceeding). • Utilities are interested in a wide range of project sizes (large and small dollar value projects). • A relationship already exists. • Utilities often have access to reduced financing rates due to their financial strength.	• Not all utilities offer UESCs (the FEMP is helping utilities launch UESC programs). • The utility might have limited renewable experience and could be uncomfortable with renewable projects. • Issues could arise regarding contracts for terms of more than 10 years; 10 years is acceptable for energy efficiency but renewable energy projects usually require a longer contract to be economically feasible.

> **Case Studies**
>
> (See Appendix C)
>
> - **Camp Pendleton, North San Diego County, California**
>   PV Project, 75kW PV www.eere.energy
>
> - **Joshua Tree National Park, California**
>   PV/propane hybrid www.eere.energy.gov/femp/pdfs/26358.pdf

The UESC enabling legislation is silent on whether the agency must take title to the project (except for the U.S. Department of Defense, which must take title but with no specification as to when). It is recommended that the utility be given title to the project for at least a few years (currently 6 years) so that tax benefits can improve the economics of the project. (See Step 5)

1. Introduction: Contract or Agreement Review

This first step provides a more detailed description of the UESC, discusses whether the local utility offers one, and—if a UESC is offered—examines what agreements might be in place. It is important to review available agreement or contract options to understand the types of projects they cover. If the type of project being considered is not covered under an existing agreement, then an agreement can be developed with the utility.

The three UESC contract or agreement types are the area-wide contract (AWC), the basic ordering agreement (BOA), and the master agreement. An AWC is a blanket contract that the GSA establishes with utilities that permits federal agencies to place orders with a utility (if contracted) for services offered under the AWC. A basic ordering agreement is made between an agency and a utility and establishes general terms and conditions for future contracts. Model agreements are template agreements developed for federal agencies, and contain required clauses for federal contracts. Template agreements can be used in whole or in part as the basis for an AWC or a BOA, or can be used alone to form a master agreement between the agency and utility. A list of area-wide contracts currently in place and a list of master agreements can be found on the FEMP Web site (www.eere.energy uescs_types.html; accessed June 10, 2010). It is important to understand that if none of the local utilities offers an UESC, then the FEMP and GSA have resources to inform the utility about UESCs (with the goal of urging the utility to offer one).

If an area is serviced by more than one utility, it is recommended that "fair consideration" be given to all utilities servicing the area. When performing a fair consideration of a utility, request descriptions of its capabilities and experience, references, and a disclosure of its markups on projects of this type. If considering a very complex project (greater than $10,000,000), the agency could choose to request more information from the utilities to help make a choice between them. After selected utilities express interest in the project, execute a justification and approval (J&A) document. It should be noted that UESCs do not have a traditional RFP process unlike the agency funded or PPA procurement mechanisms.

The FEMP can provide expert help in working through the UESC. The FEMP project facilitators and federal financing specialists have expertise in these types of projects and with alternative financing mechanisms. For more information on types of UESCs, see www.eere. energy (accessed June 7, 2010).

2. Perform a Preliminary Study

A preliminary study essentially is a walkthrough of the facility. The walkthrough constitutes a preliminary audit that is seeking energy- efficiency and renewable energy opportunities. The result of the audit is a report that outlines possible project scope and preliminary economics. This chapter is used as the basis of the decision to proceed with the project's next steps.

3. Perform an Agency Review

An agency review examines the preliminary study report. The review should look at the planned SOW and confirm that all projects that are deemed necessary (e.g., failing equipment replacement, solar project) are included in the scope. Additionally, economics and term of contract should be considered in the decision to move forward with the next step.

4. Project Implementation Proposal (Detailed Feasibility Study)

The two main components of this step are performing an IGA and drafting a project proposal based on that IGA. The IGA can be paid for up front or rolled into the financing of the overall project. The proposal provides the project scope and description, costs, schedule, and other pertinent information.

TILITY ENERGY SERVICES CONTRACT CASE STUDY

Camp Pendleton (U.S. Marines)

Naval Facilities Engineering Command (NAVFAC SW) Southwest has a vibrant and effective energy program. The team has used many tools to accomplish energy efficiency and renewable energy objectives, of which the utility energy services contract (UESC) is one. At Camp Pendleton, the phased approach started with a low cost, minimal design, energy efficiency effort that also financed the costs of audits and design for their next energy project. Using this phased approach in combination with UESCs, Camp Pendleton has installed numerous solar projects including solar thermal on five pools and a 75 kW rooftop PV system that generates 116,000 kWh/year. The phase that included the rooftop PV bundled with energy efficiency resulted in a total project cost of $11.2 million and a savings-to-investment ratio of 1.94. The contract term for the phase is 10 years. The system has been online since July 2008.

5. Negotiate and Accept the Proposal

If the proposal meets all of the needs for the project, then it either can be accepted as is or changes can be negotiated. It is anticipated that the agency-review step will address expectations of which items are to be included in the proposal, but it's important to confirm that all of the required items are incorporated. Any critical design requirements also should be checked and confirmed (e.g., design requirements regarding a rooftop PV system). Cost information should be checked and confirmed with experts on the team, and be negotiated if that is deemed necessary. Include buy down, prepayment, and termination formulas, as they can reduce finance costs and alleviate future contract administration problems. After achievement of a mutually acceptable proposal, authorization to award the contract must be received. Typically, a work release letter (WRL) or purchase order (PO) that references the requirements of the master contract and the specifics of this proposal is drafted and signed.

Another major contract consideration is determining who takes title to the renewable energy portion of the project. The UESC enabling legislation is silent on whether the agency must take title to the project (except for the U.S. Department of Defense, which must take title but with no specification as to when). It is recommended that the utility be given title to the project for at least a few years (currently 6 years) so that tax benefits can improve the economics of the project.

6. Detailed Design Acceptance

After the contract is awarded the project design phase begins. The design parameters that the system designer will work within should be clear from the accepted project proposal. The design kickoff should confirm these design parameters for all parties. It is recommended that a third-party, qualified solar-design expert (for the renewable energy portion of the project) conduct a design review at 25%, 50%, and 100% of completion, to confirm that site requirements are met. A thorough design review always is faster and less expensive than fixing design flaws later. To help with the design reviews, a "Photovoltaic Project Design Evaluation Checklist" is provided in Appendix B.

7. Construct the Project

During project construction the primary considerations are regarding coordinating with the developer. To enable successful coordination during this phase, first identify a single point of contact. Major areas of coordination are timing of work (particularly if the construction could interfere with the site's mission), ensuring that critical deadlines are met (especially those regarding incentives), assisting with interconnection issues, and handling incentive applications. The final piece of the construction process is the commissioning of the system—which makes possible the system interconnection and start up.

8. Construction Acceptance

When the system is significantly complete and operational, it is recommended that it be commissioned by a third-party expert. This is where good system design and performance specifications help ensure good value. A top-notch commissioning makes certain that the system has been installed properly and is operating to specifications. It also confirms that there are no apparent safety issues due to poor installation (e.g., damaged wire insulation, unprotected high-voltage connections). It is recommended that a good commissioning plan be established and agreed upon during the proposal process. A "Photovoltaic Commissioning Checklist" is provided in Appendix B.

Utility energy service contracts are not required to include M&V of performance and do not have to guarantee performance (as is the case for ESPCs). It is important, however, to have a plan in place that ensures that the project continues to deliver the expected savings.

9. Monitor the Performance Period

In addition to making monthly or annual payments, other concerns center on how the system operates over time and what recourse is available if it ceases to meet expectations.

Utility energy service contracts are not required to include M&V of performance and do not have to guarantee performance (as is the case for ESPCs). It is important, however, to have a plan in place that ensures that the project continues to deliver the expected savings.

One option to help ensure system performance is to include performance verification as part of the contract. Performance verification should extend for a specified period after commissioning, and should confirm that the verified performance meets a predetermined threshold. An example of this is a PV system that requires quarterly performance verification for the first year of service, and an agreement that mandates that the system output be at least 80% of calculated output based on actual solar insolation for the period. The project team and solar expert can create a reasonable agreement with the developer for a guarantee of this type, if desired. The FEMP also offers guidance and recommendations for performance assurance on page 134 of *Utility Energy Service Contract: Enabling Documents*, available at www.eere.energy documents09.pdf (accessed June 7, 2010).

If the utility agrees to long-term monitoring of the system, then the DOE Solar Energy Technologies Program can track the performance and reliability of the system installation. If interested in having the DOE track a specific system, contact Michael Quintana at Sandia National Laboratories (maquint@sandia.gov).

10. Close Out the Project

After all the payments have been made and the project term expires, either the renewable energy system is the agency's to own and maintain or the utility owns the system. Note that the DOD, however, is required to take title at some point, and civilian authority is silent on the subject of ownership. Even when the agency owns the system, performance- assurance measures still are important to the project and should be continued. If the utility owns the system, then the original contract language governs the options for what can happen to the system. Options include buying the system for fair market value, continuing to have the utility operate the system, removing the system, and abandoning the system in place. Other options that are legal under legislative authorities and acceptable to parties involved also can be considered.

ENHANCED USE LEASE

In the federal sector, enhanced use leases (EULs) have a history of being used to implement infrastructure building projects. Now, they are also being used to realize solar energy projects. An EUL is a real estate agreement that focuses on underutilized land. Prospective developers compete for the lease, and payment can be either monetary or in-kind consideration (in this case, renewable power can be part of the consideration). The value of the lease is used to determine the amount of consideration. An EUL typically is used for large projects, for example those having a capacity that is greater than the site load. A few agencies have the authority to execute an EUL.

Steps to Follow

1. Identify underutilized land
2. Develop the scope of work
3. Develop a notice of opportunity to lease
4. Issue a notice of opportunity to lease
5. Administer the notice of opportunity to lease
6. Evaluate proposals
7. Select a contractor
8. Complete a lease and management plan (contractor submittal)
9. Review and accept the plan
10. Issue the lease
11. Design the project
12. Construct the project
13. Commission the project
14. Monitor the performance period
15. Perform project close out

Pros	Cons
• Discovers unrealized value of underutilized property. • Supplements underfunded facilities costs. • Can be used in combination with the ESPC, UESC, and PPA.	• Currently only the DOD, the DOE, National Aeronautics and Space Administration (NASA), and the U.S. Department of Veterans Affairs (VA) have the authority to execute an EUL. • Must not be excess property as defined by 40 U.S.C. § 102.

Case Studies

(See Appendix C)

- **NASA Kennedy Space Center/ Florida Power & Light Company, Merritt Island, Florida**
    10-MW PV, owned by Florida Power & Light Company with in kind consideration of a 990-kW NASA-owned PV
    www.smartgridnews.com/artman/uploads/1/nasa_space_coast_solar.pdf
    www.fpl.com/environment/solar/ spacecoast.shtml
- **Fort Irwin (in process), Barstow, California**
    500-MW solar thermal/PV by 2022
    http://eul.army.mil/ftirwin/

1. Identify Underutilized Land

The prerequisite for an EUL is that the site has "underutilized" land that is not "excess property" as defined in 40 U.S.C. § 102. Identify land that meets the "underutilized" requirement and that is a reasonable site for a solar project. A market appraisal must be completed to determine land value.

2. Develop the Scope of Work

The SOW becomes the basis of the notice of opportunity to lease (NOL) used to solicit project proposals from solar developers. This is a statement of all the requirements of the project. The solar feasibility study should provide the basis of the project's SOW. The team's solar expert is a key person in the development of this document.

The SOW should include:

- Type of renewable energy systems required,
- Expected size of systems,
- Location on-site, and
- Specific site requirements.

3. Develop a Notice of Opportunity to Lease

An NOL is the document issued to the public to solicit offers from solar developers. This document describes the content to be included in proposals and provides relevant information that developers can use when a making an offer. The NOL should include the following elements.

- Clarification of which party is responsible for procuring permits and arranging contracts for energy generation offtakers.
- Commissioning plan.
- Criteria used to evaluate proposals.
- Description of how the proposal process is to be administered (e.g., proposal meetings, site visits, process for answering questions).
- Due diligence.
- Limits on proposed project timeline.
- Post-commissioning performance.
- Proposed financing structure and financial industry commitments.
- Scope of work.
- Timelines for proposal process.

The description of the proposal process administration includes the timing and location of industry forums and site visits, and the process to be used for answering questions. The industry forum provides the opportunity to present the project requirements in detail to interested developers, and for developers to ask questions. The site visit allows interested developers to assess site conditions and ask additional questions. The site visit can be held in conjunction with the industry forum or can be conducted separately. Depending on the site conditions and the process chosen, additional site visits could be necessary for respondents to perform additional due diligence.

Questions that arise during and after the industry forum and site visit must be handled such that all developers have access to the same information. There are many variations on how this can be accomplished. A recommended approach is to write down every question, answer each, and post the questions and answers on a Web site. This process can help to avoid variation in answers, participant misunderstandings, and the potential for one party to receive more information than another.

Describing the criteria and the process to be used to evaluate proposals helps developers structure their responses and ensures that all primary issues are addressed. This also makes it easier to review proposals. The respondents address identified criteria in separate sections, eliminating the need for a reviewer to pick statements from the entire proposal that apply to the criteria. Evaluation criteria should include the considerations that are most important to the specific project, and could include the following.

- In-kind consideration.
- Developer's experience, performance track record, and references.
- Developer's financial health.
- Developer's ongoing long-term service capacity.
- How specific site issues and requirements will be addressed.
- Amount of energy generated.
- Quality of components.
- Local sourcing of components and labor.

Describing how to handle due diligence on the developer's part has important contractual implications. Due diligence is the effort the developer must put forth to fully understand the project and the risk of any unknowns that could arise. Contractually, the developer wants to have recourse if something unexpected comes up that is outside its ability to perform or that will cause significant cost increases (such as roof structural issues or subterranean rock that would preclude standard trenching machines). Under such circumstances, the developer might want to be able to walk away from the project or have the option to renegotiate. Typically, developers should factor the risk into their proposals and be required to deal with the unexpected.

Depending on the project, due diligence could require considerable effort and expense. If this is the case with a project, or if this becomes apparent as the NOL process progresses, consider adding additional steps to the NOL process. Additional steps can include an initial proposal review to generate a short list of developers. Those on the short list are invited to continue with the NOL process, which includes expending additional effort and incurring additional expense for due diligence. The purpose of this extra step is to assure the developers on the short list that they have a good chance at being successful, and that it is worthwhile to put forth the extra effort and expense required for due diligence. A site due diligence date— after which the developer will have entered into an irrevocable contract— should be specified. An example of a NOL can be found at http://eul.army.mil/ftirwin/Docs/FinalNOL20Mar09.pdf (accessed June 8, 2010).

4. Issue the Notice of Opportunity to Lease

After the NOL is complete, announce it somewhere that developers can find it. One prominent Web site that is familiar to most developers is www.fbo.gov (accessed June 8, 2010). Another Web site option is the Green Power Network, http://apps3.eere.energy.gov/greenpower/financial/ (accessed June 8, 2010). It should be noted that EULs do not have a traditional RFP process unlike the agency funded or PPA procurement mechanisms.

5. Administer the Notice of Opportunity to Lease

After the NOL has been issued, follow the defined timeline and described NOL process. Adjust it as necessary if unforeseen events arise. This step may include site visits, pre-bid meetings, and correspondence related to project questions and answers.

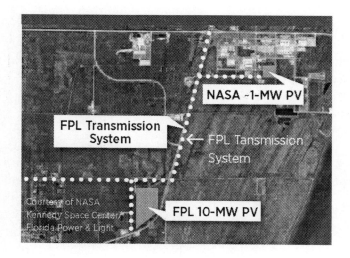

ENHANCED USE LEASE PROJECT CASE STUDY

Kennedy Space Center (NASA)

Operation and mission changes at the Kennedy Space Center led to the search for expanded business opportunities. Capitalizing on the vast amount of land on Florida's Space Coast, NASA's history and experience with renewable energy, and NASA's strong partnership with Florida Power and Light (FPL), renewable energy projects became a focus of investigation. Wind energy was the first choice due to cost and power density, but issues with the resident wildlife refuge put wind projects on hold. Photovoltaics became the preferred project type. NASA and FPL agreed on appraised value, lease terms, and two sites totaling 60 acres for development in phase 1. The thirty-year lease is for a site where FPL will construct, own, and operate a 10MW PV plant that will feed power directly into the FPL transmission system. The $6,300,000 consideration for this lease is a NASA-owned 990kW PV plant that will feed into the NASA-owned distributions system but that will be constructed and maintained by FPL.

6. Evaluate the Proposals

A small team should be assembled to evaluate the offers received. The number of team members to include depends upon the specific project, but the team should have at least three people. Most of the people on this team probably will come from the project team. Other key people to consider including on this team are the:

- Energy manager,
- Facilities manager,
- Project manager,
- Site manager (if managers for different areas of a site), and
- Solar expert.

The process for evaluating the offers should have been established during the development of the NOL. It is recommended that the proposal-evaluation sessions be scheduled well in advance to ensure the availability of key personnel. These review sessions typically take a week, unless an unusually great number of proposals (more than 10) is received. Follow the proposal evaluation criteria described in the NOL and, from the start, clearly define the meaning of each criteria and score.

7. Select the Contractor

The proposal evaluation scores the offers. This helps determine the most advantageous offer and, consequently, which contractor to select.

8. Complete a Lease and Management Plan

Once selected, the contractor and agency initiate steps toward entering into a lease and management plan (L&MP). The L&MP describes what the contractor will do and how elements will be completed. An L&MP can include the following components.

- Architecture and engineering.
- Community and stakeholder relations.
- Consultation and coordination in accordance with the Section 106 process as required by the National Historic Preservation Act (NHPA) (if applicable).
- Consultation and coordination with federally recognized Native American tribal governments (if applicable).
- Cultural or archeological documentation.

- Financial.
- Legal services.
- Master planning.
- Marketing.
- NEPA and other environmental documentation.
- Site assessment.

The L&MP is a significant piece of work, and a minimum of 18 months elapses from the time a contractor is selected to the completion of a lease. Prior to completion of L&MP, all NEPA compliance analysis addressing proposed project impacts on land within the lease must be completed, through either environmental assessments or environmental impacts studies. The contractor submits the completed L&MP to the agency.

9. Review and Accept the Plan

The completed lease and management plan are reviewed by the agency. If there are any unresolved issues, then last negotiations are completed and the L&MP is accepted.

10 Issue the Lease

After the lease and management plan is accepted, a lease can be finalized and issued. A sample EUL contract can be found at www.eere.energy (accessed July 19, 2010).

11. Design the Project

After the contract is awarded, the project final design phase (based on preliminary design work in the L&MP) begins. The design kickoff should confirm the design parameters for all parties. It is recommended that a third-party, qualified solar-design expert conduct reviews at the 25%, 50%, and 100% design completion, to confirm that site requirements are met. A thorough design review always is faster and less expensive than fixing design flaws later.

An EUL is a real estate agreement that focuses on underutilized land. Prospective developers compete for the lease, and payment can be either monetary or in-kind consideration (in this case, renewable power can be part of the consideration).

12. Construct the System

During project construction the primary considerations are regarding coordinating with the developer. To enable a successful coordination during this phase, first identify a single point of contact. Major areas of coordination include the timing of work (particularly if construction could interfere with the site's mission), ensuring that critical deadlines are met (especially those regarding incentives), assisting with interconnection issues (including interconnection and net metering agreements), and handling incentive applications. The final piece of the construction process is the commissioning of the system—which makes possible the system interconnection and start up.

13. Commission the System

When the system is significantly complete and operational, it is recommended that it be commissioned by a third-party expert. This is an important step, as this system is located on the agency's site and system performance can affect the in-kind consideration. A good commissioning ensures that the system has been installed properly and is operating to specifications. It also confirms that there are no apparent safety issues due to poor installation (e.g., damaged wire insulation, unprotected high-voltage connections).

14. Monitor the Performance Period

Operation and maintenance of the system is the responsibility of the developer. If in-kind consideration is dependent on the system's energy production, then a process for monitoring energy production should be in place. If the developer owns the RECs produced by the project and the agency must meet renewable energy targets, replacement RECs can be purchased. These RECs can count twice toward the agency's targets as long as the project produces energy at levels greater than or equal to the quantity of RECs purchased.

If the developer agrees to long-term monitoring of the system, then the DOE Solar Energy Technologies Program can track performance and reliability of system installations. If interested in having the DOE track a specific system, contact Michael Quintana at Sandia National Laboratories (maquint@sandia.gov).

An EUL typically is used for large projects, for example those having a capacity that is greater than the site load. A few agencies have the authority to execute an EUL.

15. Close Out the Project

The end of the lease options and decisions are characterized by the economic value and viability of the installed system. If the system still can be operated economically then several options can be considered, including extending the lease, purchasing the system for fair market value, or entering into a PPA-type agreement with the developer. If the system has limited or no economic viability at the end of the lease then it can be abandoned in place or removed. This choice most likely will be influenced by circumstances existing 20 to 75 years in the future, and this should be taken into account when developing the options at the start of the project.

GLOSSARY AND RELATED SOLAR TERMINOLOGY

A

Agency-funded project The project is funded through an agency appropriation and the agency owns the project.

Alternating current (AC) The movement of electric charge periodically reverses direction. AC is the form of electric power delivered to businesses and residences.

Alternative compliance payment In lieu of standard means of compliance with renewable portfolio standards, electricity suppliers may make alternative compliance payments to make up for deficiencies (in megawatt-hours) between the amount of electricity from renewable resources mandated and the amount actually supplied. Payment amount varies among states.

Array A group of photovoltaic (PV) modules (also called solar panels) or solar thermal collectors.

Authority having jurisdiction (AHJ) A federal, state, or local entity having statutory authority for approving equipment, an installation, or a procedure.

Avoided-cost rate The cost per kilowatt-hour a utility would have incurred by supplying electricity generated from its traditional generation sources.

B

Behind the meter Refers to the location where a generating technology (such as a PV system) is connected to the electricity grid. A behind-the-meter PV system is connected between the utility meter and the facility using the electricity, so all electricity generated by the PV system that is not being used by the facility flows through the utility meter to the grid.

Binomial tariff A utility rate structure that includes both a fixed demand charge and a variable (per kilowatt-hour) energy charge.

British thermal unit (Btu) The amount of heat required to raise the temperature of one pound of water from 60°F to 61°F at a constant pressure of one atmosphere. Water heating is commonly measured in Btus.

Building energy code Establishes minimum energy performance features in buildings.

Building integrated PV (BIPV) Standard PV modules, transparent modules, and thin-film covers and tiles are used to replace or enhance conventional building materials such as roofs, walls, facades, awnings, and skylights. These materials generate electricity from sunlight and perform other functions integral to the building's design.

Building integrated solar water heating (BISWH) Similar to BIPV, BISWH incorporates solar water heating materials into traditional building materials.

Buy-down A reduction in costs to purchasers.

C

Capacity limit for individual systems A limit placed on the capacity of individual PV systems, usually set to a certain percentage (for example, 125%) of a customer's energy load. Capacity limits can vary by utility type, solar energy system type, or customer type.

Carbon dioxide (CO_2) A colorless, odorless, noncombustible gas present in the atmosphere. It is formed by the combustion of carbon and carbon compounds (such as fossil fuels and biomass); by respiration, which is a slow combustion in animals and plants; and by the gradual oxidation of organic matter in the soil. Considered a greenhouse gas that contributes to global warming. See also *emissions*.

Clean renewable energy bond (CREB) Special-purpose tax credit bonds that provide the equivalent of an interest-free loan for certain qualifying energy facilities. Bondholders receive a tax credit on their federal income taxes instead of an interest payment from the bond issuer.

Credit multiplier A credit multiplier for solar offers additional credit toward compliance with a renewable portfolio standard for energy derived from solar resources.

Code official Local government employee who enforces codes and standards, ensuring that solar energy system installations meet applicable safety, building, electrical, and plumbing codes in a region.

Commercial energy conservation ordinance (CECO) A CECO requires commercial property owners to complete certain energy conservation measures in their buildings upon transfer of property ownership or when additions or renovations are made.

Customer aggregation program A program that coordinates group purchases of solar energy systems, helping defray some of the up-front costs of solar installations by giving aggregated individuals or businesses a discounted rate for bulk purchases of solar energy systems.

Customer generator Utility customer who generates electricity on his or her property using a distributed generation technology such as PV.

Customer-sited distributed generation Refers to distributed generation technologies such as PV installed on the property of a utility customer.

D

Demand charge A charge incurred by a utility customer in return for the utility having built adequate generating capacity to supply the power needed for a facility (like a manufacturing plant) to operate at its maximum capacity.

Design development Takes the preferred alternative from Schematic Design and develops the details such as the transport phenomenon (wire and pipe sizes), and assembles schedules of the required equipment.

Direct current (DC) The unidirectional flow of electric charge. DC is the form of electric power produced by photovoltaics and batteries.

Direct incentive Cash back to consumers for a qualified solar installation. Direct incentives include up-front rebates and grants and production-based incentives that are typically distributed over several years.

Distributed generation Electricity production that occurs on site (or close to the load center) and is interconnected to the utility's electric distribution system.

Dollars per watt ($/Watt) The standard metric for assessing the cost of a solar electric system. This metric is either the total installed cost of a system (or a component of the system) divided by the name plate capacity of the system (or inverter, if looking only at the inverter).

Due diligence The work a developer needs to do to ensure it has assessed a project and its risks to the extent necessary to commit to a binding contract.

E

Electric capacity The amount of electricity-generating resources a utility must supply to meet the demands of a particular facility or region.

Electric utility A corporation, agency, authority, or other legal entity aligned with distribution facilities for delivery of electric energy for use primarily by the public. Investor-owned electric utilities, municipal and state utilities, federal electric utilities, independent system operators, and rural electric cooperatives are included.

Electricity distribution system The portion of the electricity grid that distributes lower voltage electricity from high-voltage transmission lines to individual homes and businesses.

Emissions In the context of global climate change, emissions refer to a release of greenhouse gases into the atmosphere, such as CO_2, methane, and oxides of nitrogen.

Energy audit A survey that determines how much energy is used in buildings and facilities, which helps identify ways to use less energy.

Energy conservation mechanism (ECM) A training program, facility improvement, or equipment purchase used to reduce energy or operating costs in a building.

Energy savings performance contract (ESPC) An agreement between a building owner (or facilities manager) and a private energy services company (ESCO) that uses future energy savings to pay for the entire cost of a building's electricity and energy efficiency retrofits.

Energy service agreement (ESA) An agreement contained within an energy savings performance contract (ESPC) that addresses the assignment of ownership of a system to allow a private party to benefit from tax incentives afforded the renewable energy system.

Energy services company (ESCO) A company that offers energy management services to reduce a client's utility costs. Cost savings are often split with the client through an energy performance contract or a shared- savings agreement.

Enhanced use lease (EUL) A real estate agreement that leases "underutilized land" to a developer for cash or in-kind consideration.

Environmental justice The fair treatment and meaningful involvement of all people, regardless of race, color, national origin, or income, with respect to the development, implementation, and enforcement of environmental laws, regulations, and policies.

Escalator Regarding solar projects, this is the rate an energy price will increase annually. An escalator can be part of a power purchase agreement (PPA) price, which means that the PPA price for energy increases by the escalator rate every year.

Expected performance rebate Cash incentive based on the expected energy output from a solar energy system over a given period of time.

External utility-accessible AC disconnect switch A hardware feature that allows a utility employee to manually disconnect a customer-owned PV system (or other type of generation) from the electricity grid.

F

Feasibility study Engineering and economic viablity of a project are evaluated in a feasibility study. Engineering viablilty considers the physical equipment and connections to existing infrastructure to determine if a system will work physically. Economic feasibilty estimates the costs and revenues of a project and applies economic theory to determine if a project will be financially viable. A feasibilty study must also include issues such as environmental constraints (endangered species, etc), historic preservation, or any other factors which may delay or impede implementation.

Federal investment tax credit A credit against federal income taxes, usually computed as a percentage of the cost of investment in solar energy assets. The federal investment tax credit for installing solar energy systems is set at 30% of the installed system cost, and is set to expire in 2016.

Feed-in tariff (FIT) A renewable energy policy that typically offers renewable energy project developers a guaranteed payment for electricity produced by their renewable energy system over a fixed amount of time (usually 15 to 20 years).

G

General fund The primary operating fund of a governmental entity, usually in place to support operating expenditures.

Generating capacity The amount of power-generating resources a utility can supply to meet the demands of a particular facility or region.

Gigawatt (GW) A unit of power equal to 1 billion watts, 1 million kilowatts, or 1,000 megawatts.

Green pricing A mechanism for utility customers to support their utility's investments in renewable energy projects through direct charges on their monthly utility bills. Green pricing is a market-based solution to account for the nonmarket (meaning environmental) benefits of renewable energy.

Greenhouse gas Atmospheric gasses that absorb and emit radiation. Greenhouse gases as defined by E.O. 13514 include carbon dioxide, methane, nitrous oxide, hydrofluorocarbons, perfluorocarbons, and sulfur hexafluoride.

I

IEEE 1547 IEEE was originally an acronym for the Institute of Electrical and Electronics Engineers, Inc. Today, the organization's scope has expanded into so many related fields that it is simply referred to by the letters IEEE ("I-triple-E"). IEEE 1547 refers to the *Standard for Interconnecting Distributed Resources with Electric Power Systems*.

Indefinite-delivery/indefinite-quantity (IDIQ) contract A base contract with preapproved suppliers that accepted the contract terms. Meant to streamline the contracting process, as only project specific contractual issues need to be negotiated.

Installation baseline An accounting of all solar energy installations currently in place.

Installation target A set goal for future solar installations in an agency by a specific date. A solar installation target is often set to achieve broader environmental, climate, or sustainability goals.

Installed capacity The total amount (usually measured in terms of size; in kilowatts or megawatts for PV) of solar energy systems operating in a given region or sector.

Interconnection The process of connecting an electricity-producing technology (like a PV system) to the electricity grid.

Interconnection agreement Agreement between a utility and a customer that specifies the terms and conditions under which solar electric systems or other approved customer-owned generation will be connected and operated.

Interconnection standard A technical, legal, and procedural requirement that customers and utilities must abide by when a customer wishes to connect a PV system to the grid.

Investment Tax Credit A tax incentive based on the initial cost (investment) of an energy system.

K

Kilowatt (kW) A standard unit of electrical power equal to 1,000 watts.

Kilowatt-hour (kWh) A unit of energy: 1,000 watts acting over one hour.

L

Leadership in Energy and Environmental Design (LEED) LEED is a voluntary, consensus-based national rating system for developing high-performance, sustainable buildings operated by the U.S. Green Building Council.

Levelized cost (of energy) (LCOE) A means of calculating the cost of generating energy (usually electricity) from a particular system that allows one to compare the cost of energy across technologies. LCOE takes into consideration the installed solar energy system price and associated costs such as the cost of financing, land, insurance, operation and maintenance, and other expenses.

Life cycle cost The present value of all costs associated with purchasing and operating an asset (solar system) during its expected life.

Liquidated damages (LD) These are typically the financial penalties assessed on developers who fail to meet certain deliverable stage gates in an established timeframe or fail to complete the project at all or in the established timeframes.

Load Describes the amount of power (amps) consumed by an electrical circuit or device. Loads are usually expressed in amps but sometimes in watts.

M

Megawatt (MW) Standard measure of electric power plant generating capacity equal to 1,000 kW or 1 million watts.

Megawatt-hour (MWh) 1,000 kWh or 1 million watt-hours.

Metric ton of carbon dioxide equivalent (MtCO$_2$e) Standard measurement of the amount of CO_2 emissions reduced or secluded from the environment.

N

National Environmental Policy Act (NEPA) This act requires all projects implemented on federal property or built with federal funding to follow a process that considers the environmental impacts and mitigation measures, if necessary, for the project.

Net metering Net metering is a billing mechanism that credits solar system owners for the electricity exported onto the electricity grid. Under the simplest implementation of net metering, a utility customer's billing meter runs backward as solar electricity is generated and exported to the electricity grid and forward as electricity is consumed from the grid.

P

Peak sun hours The equivalent number of hours per day when solar irradiance averages 1,000 watts per square meter.

Permitting incentive Incentive that reduces or waives local permit fees, plan check fees, design review fees, or other such charges consumers and businesses may incur when installing a solar energy system.

Photovoltaic (PV) system A set of components for converting sunlight into electricity. Comprises the solar modules or array that captures the sunlight along with balance-of-system

(BOS) components, such as the array supports, electrical conductors/wiring, fuses, safety disconnects and grounds, charge controllers, inverters, and battery storage.

Power purchase agreement (PPA) A legal contract between an electricity generator and electricity purchaser. Solar PPAs typically provide a long-term contract to purchase electricity generated from a solar installation on public or private property; a type of third party ownership model.

Procurement specification Procurement specifications describe the requirements of a project so that offerors may prepare a bid. Specifications often include an introductory background; scope of work, deliverables, and schedule. Specifications include the goals of the project, specified performance requirements, quality control requirements, interface requirements (how the new system touches existing systems), scheduling and access requirements, reporting requirements, and any other restrictions on how the offereror may complete the scope of work.

Production-based (or performance-based) incentive Cash payment to project owners based on electricity production on a dollar-per-kilowatthour basis over a specified duration.

Production tax credit A tax incentive based on the MWh production of an energy system.

Project developer A company that provides services for solar installations including planning, organizing, executing, and managing resources for installation projects.

Public benefits fund A fund dedicated to supporting renewable energy and energy efficiency projects. The fund is typically financed through a small charge on the bill of utility customers (sometimes referred to as a system benefits charge) or through specified contributions from utilities, although other means of funding such as legislative appropriations are possible.

R

Radial electric distribution system The dominant electric distribution system in the United States where electricity is supplied from a single source and there are no closed "loops" in the system.

Real-time pricing (RTP) The instantaneous pricing of electricity based on the cost of the electricity at the time it is used by a utility customer. RTP rates are volatile and are generally very high when demand for electricity is high.

Rebate Cash incentive issued to a purchaser of a solar energy system to help defray the up-front cost of installing the system.

Renewable energy (RE) Energy from resources that naturally replenish themselves and are virtually inexhaustible. Renewable energy resources include biomass, hydropower, geothermal, solar, wind, ocean thermal, wave action, and tidal action.

Renewable energy certificate or credit (REC) A REC represents the property rights to the environmental, social, and other non-power qualities of renewable electricity generation. A REC, and its associated attributes and benefits, can be sold separately from the underlying physical electricity associated with a renewable- based generation source.

Renewable energy certificate (REC) marketer or aggregator A REC marketer or aggregator buys RECs at wholesale prices and sells RECs at retail, similar to a commodities dealer.

Renewable energy certificate (REC) trading mechanism An exchange for trading RECs much like the New York Stock Exchange for trading shares in companies.

Renewable portfolio standard (RPS) A mandate requiring that renewable energy provides a certain percentage of total energy generation. The mandate is sometimes referred to as a renewable electricity standard or RES.

Request for Information (RFI) A document that requests feedback from interested parties on the best approach to development on sites. Information gathered from this process is typically used to inform development of an RFP.

Request for proposal (RFP) A document issued to the public to solicit project proposals from solar developers.

Request for qualifications (RFQ) A publically issued request for statements of qualification from solar developers that are interested in bidding on a solar project. The purpose is to obtain a pre-qualified list of developers that will be allowed to respond to the RFP.

Revolving loan fund A source of money from which loans are made. As loans are repaid, funds become available for new loans to other entities.

S

Sales tax incentive Exemption from or refund of sales tax for purchasing and installing solar energy components and systems.

Schematic design Schematic design describes the size and type of major components, and the relationships between those components. Schematic design should answer all questions related to thermodynamics (energy balance) but not necessarily related to transport phenomenon (wire and pipe sizes). Issues of physical size and location should be addressed.

Schematic design often involves evaluation of alternatives and selection of a preferred alternative.

Screening Screening is a preliminary analysis based on existing (or easily collected) data to determine which sites are candidates for a feasibility study. Screening is often based on high-level information such as utility bills and published climate data. Screening is most often conducted on a large number of sites to focus an agency's effort on sites most likely to have viable projects.

Secondary network distribution system A type of electric distribution system that serves central business districts in many cities. These systems contain multiple feeders and transformers to provide excellent service reliability and the capacity to serve large loads, such as high-rise buildings.

Service entrance capacity The amount of power a building is designed to handle. A service entrance is the point at which electricity enters a building. A service entrance switchboard has metering equipment and devices for overcurrent protection and electrical control.

Set-aside A mandate or goal for some fraction of a renewable portfolio standard to be met with designated technologies such as PV.

Solar access The ability of one property or area to continue to receive sunlight without obstruction from a nearby home or building, landscaping, or other impediment.

Solar aggregation purchasing program See customer aggregation programs.

Solar bulk purchasing See customer aggregation programs.

Solar Decathlon An international competition between colleges and universities in which teams compete to design, build, and operate the most attractive, effective, and energy-efficient solar-powered house. The competition, which is sponsored by the U.S. Department of Energy, takes place every two years in Washington, D.C.

Solar easement A type of solar access law that grants the owners of solar energy systems the right to continued access to sunlight without obstruction from a neighbor's property and limits future property developments that could restrict solar access.

Solar electricity See photovoltaic system.

Solar energy Electromagnetic energy transmitted from the sun (solar radiation). The amount that reaches the earth is equal to one billionth of total solar energy generated or the equivalent of about 420 trillion kWh.

Solar farm Refers to a large-scale solar installation.

Solar installer licensing Licensing requiring a baseline of quality below which it is illegal to operate.

Solar permitting process To install a grid-connected PV system, the homeowner or builder must obtain an electrical permit and in some cases a building permit from the local government, followed by an inspection of the installation. Solar water heating systems require a plumbing permit and sometimes a building or mechanical permit, or both.

Solar-ready A solar-ready home or building is designed as if a solar energy system were going to be installed during construction. Architects and builders take precautions to ensure a viable site for solar technologies by leaving adequate roof space free from vents, chimneys, and equipment; planning landscaping to avoid shading the unobstructed roof space in the future; planning extra space for equipment in mechanical rooms; preinstalling roof mounting systems and conduit; and labeling structural reinforcements and end points of wires or pipes.

Solar resource The amount of sunlight a site receives, usually measured in kilowatt-hours per square meter per day. See also peak sun hours.

Solar right law A law or ordinance that furnishes protection for homes and businesses by limiting or prohibiting restrictions (for example, neighborhood covenants and bylaws, local government ordinances, and building codes) on the installation of solar energy systems.

Solar site assessment An evaluation of a site being considered for a solar energy installation. A trained solar site assessor collects data such as roof or property orientation and slope, dimensions of available installation space, electrical and/or plumbing configuration, and shading on the site location.

Solar thermal Solar energy conversion technologies that convert solar energy to thermal energy (heat) used to heat water or provide energy for space heating and cooling in active solar space heating or cooling systems.

Spot-market A market in which commodities are bought and sold for immediate delivery.

Stub-out The result of preparing a building for future equipment installations. To prepare for solar electric systems, conduits are run through the building so wires can connect a PV system to an electrical panel at a future date. For solar water heating systems, open-ended pipes are placed in an accessible location to connect solar collectors to hot water storage in the future.

Sustainable solar infrastructure The social, economic, policy, and physical networks and institutions that enable solar energy to be used as a mainstream energy source even in the absence of significant government subsidies.

System benefits charge A small charge on the bill of utility customers to support public policy initiatives such as renewable energy and energy efficiency programs.

System capacity The maximum expected energy production from a PV system.

System rating A rating of the maximum power a solar energy system will produce under standard test conditions (STCs). STCs are a solar irradiance of 1,000 watts per square meter, a temperature of 77°F, and an air mass of 1.5. Solar irradiance is measured in watts per square meter of light incident on Earth.

T

Tariff A document approved by the responsible regulatory agency that lists the terms and conditions—including a schedule of prices—under which utility services will be provided.

Time-of-use (TOU) pricing (or tariff) A rate schedule in which the utility customer is charged different amounts for power based on the time of day and season. Typically, peak rates are during summer afternoons. Solar customers who generate power during peak rates are credited by the utility company at those peak rates.

Transmission and distribution loss The energy lost when transporting electricity over long distances through the electricity grid's transmission and distribution systems from central generation plants to the point of electricity consumption (homes and businesses).

True up When a utility calculates the "net" consumption versus generation over a given time period (month or year). Compensation for net excess generation is often limited to the amount of electricity used during the true-up time period. Monthly true-up cycles don't capture the true value of a PV system's generation because excess generation in the summer (when PV is producing at its peak) is lost and consumption during winter (when PV systems are producing at their minimum) is charged.

U

Utility energy services contract (UESC) A contract with a "serving" utility for the utility to provide comprehensive energy and water efficiency improvements and demand reduction services.

Utility renewable electricity service contract (URESC) A contract with a "serving" utility for the utility to provide electricity from on-site solar electricity generation. As of the summer of 2010, no URESCs had been completed.

Sources

U.S. Department of Energy. *Solar Powering Your Community: A Guide for Local Governments*. www.solaramericacities.energy

Accessed March 2009.
California Energy Commission. *Glossary of Energy Terms.*
www.energy (and references therein). Accessed March 2009.
DSIRE Glossary.
www.dsireusa.org/glossary/ Accessed March 2009.
Energy Information Administration. Glossary. www.eia.doe.gov/glossary/index.cfm Accessed March 2009.
U.S. Department of Energy. Glossary of Energy-Related Terms. www.eere.energy Accessed March 2009.

APPENDIX A: SOLAR TECHNOLOGIES OVERVIEW

Federal site managers need to be smart buyers of solar energy and not necessarily experts in solar technology. The sections below provide a brief technology overview for background information. The focus is on small to medium distributed solar systems.

The solar technologies explained are:

- Photovoltaics
- Solar hot water (both domestic hot water and pool heating)
- Solar ventilation pre-heat

Other topics related to solar energy in buildings include passive solar heating, cooling load avoidance, and daylighting. These topics are not covered here as they are usually part of a larger architectural design of a building rather than a financed project.

Photovoltaics

How Photovoltaics Work and Types of Systems

A photovoltaic (PV) system creates electricity when light shines on it, but it needs to be integrated into systems that can power electric machines, appliances, lights, and other electric devices. A PV module creates relatively low- voltage direct current (DC) electricity. PV modules are connected in series and in parallel to create arrays. Series connections increase the voltage of a string, and parallel connections increase the current delivered by the array. Typically the installer defines the maximum DC voltage that the array can produce and designs the series strings to produce close to but below that voltage. The strings are then connected in parallel to complete the PV array. For most grid-connected systems, voltage must be limited to 600 volts direct current (VDC), requiring many parallel strings for large systems. These strings are combined in a combiner box to larger dimensional wires. The exception to this design method is if micro inverters are used, which are explained below.

The three types of PV system designs are grid- connected without storage, grid-connected with emergency battery backup, and off-grid with battery storage.

Grid-connected without storage systems are by far the most common and most likely will be the type to consider for the site. Grid-connected systems without storage consist of two critical parts: PV array and utility-interactive inverter. The inverter changes the DC power and voltage the array produces into alternating current (AC) power at the voltage and number of phases appropriate for the interconnection point. The inverter also senses the utility power frequency and synchronizes the PV-produced power to that frequency. When utility power is not present, the inverter will stop producing AC power to prevent "islanding" or putting power into the grid while utility workers are trying to fix what they assume is a de-energized distribution system.

If the site is connected to the grid and has critical loads that need emergency backup power, a grid-connected with emergency battery backup system is a consideration.

If the site does not have grid power and runs on generators, an off-grid system with battery storage can reduce generator run time and save on operations and maintenance costs.

PV systems with battery storage include a charge controller that is either integrated into the inverter or a separate component. The charge controller controls both the DC voltage that comes off the PV array and the voltage going into the batteries. Batteries require the specific stages of charging produced by charge controllers to help extend their longevity.

Suitland Federal Center PV system, Maryland. (*Credit: Bob Madani for DOE/FEMP. PIX 10726.*)

All types of PV systems require switch gear and protections as directed by electrical code (e.g., NEC 690) and good system design to protect the equipment. This gear may include DC disconnect; AC disconnect; lighting surge arrestor; ground fault current interrupter; and fuses or breakers and transformers required for higher voltage interconnections. Many utilities require redundant utility-specified relays.

PV Technologies

There are presently two primary technologies that form most of the PV market: crystalline silicon and thin films. Crystalline silicon PV is constructed on wafers of silicon, while thin films are constructed by depositing thin layers on a substrate.

Crystalline modules are the most efficient in converting the sun's energy to electricity. In other words, they produce the most electrical power per area of module. Typical efficiencies for crystalline PV are between 10% and 19% (this is the percentage of the sun's energy that strikes the module and is converted to electrical energy). This type of module tends to be the most expensive per unit of power capacity (dollars per watt) compared to thin film PV technology. Crystalline can be a good match for projects that have limited space and are seeking capacity maximization or where set up costs (i.e., mounting and installation costs) are high.

Crystalline type modules have a lifespan in the 25-30 year range but can keep producing energy beyond this range. They have a proven track record and typically have 25-year power output warranties. On the downside, the modules become less efficient as their temperature rises, and even partial shading can significantly reduce the power output of the strings involved. Crystalline modules presently make up the majority of the market for installed PV.

Thin film PV technologies are the least efficient, with typical efficiencies between 4% and 12%. Some thin film products don't have the proven track records that crystalline PV enjoys and are considered to have a shorter life span, but most come with 25-year power warranties. However, they can be less expensive per watt and are less susceptible to shading and temperature effects. Thin film PV products are designed for several different installation options, but one thrust is for building-integrated installation, especially easy integration with roofs. This type of PV can be a good match for several situations: very large arrays with low set-up cost; large roofs that do not require maximization of production or are structurally insufficient to handle a heavier racked crystalline PV installation; in areas at risk for hurricanes and other extreme weather; near roads where debris can be thrown up by transportation; or ground-mounted systems where earth settlement is an issue.

Inverters

There are two types of inverters for grid- connected systems: string and micro inverters. Each type has its strengths and weakness that may recommend them for different types of installations. String inverters are most common and typically range in size from 1.5 kW to 500 kW. Benefits of these inverters are that they tend to be less expensive per watt of capacity, selection includes a large range of output voltages, and possible extended warranties may be available of up to 20 years on larger units. On the down side, if the inverter fails the project could lose a significant part of production during the outage. For larger systems, multiple inverters can be combined in parallel and still produce a single point of interconnection with the grid. Warranties typically run between 5 and 10 years, with 10 years being the current industry standard. Typically, inverters will have to be replaced during the life of the system. Micro inverters are new and are designed to be dedicated to the inversion of a single PV module's power output. Small projects with irregular modules and shading issues typically benefit from micro inverters.

Mounting Systems

PV mounting system can generally be divided into three types: building mounted, ground mounted, and canopy. The type of mounting system will depend largely on the physical site as well as possible secondary goals. If the site is an urban one with no land or parking lots, a building-mounted system will be required. If land is available on the site, considerations may include future plans for the land, possible environmental hurdles to a ground-mount system, desire for covered parking, desired system size, and access restrictions, among others.

Building-mounted PV is primarily either integrated or racked systems on the roof. Integrated systems include PV sunshades for windows, PV in walls, and roof-adhered systems, among other options. Many forms of integrated PV are directed at new construction; however, roof-adhered systems (where the PV is adhered to the roof) are an option for most buildings with a reasonably new roof or those scheduled for a new roof. A roof-adhered system is a thin film PV module that is adhered to a roof surface, ranging from a membrane to a metal roof. This system can be an option when a roof's weight-bearing capacity or aesthetics are an issue.

Roof-mounted racking systems are designed for framed PV modules. The two classifications of roof racking systems are ballasted and attached (roof penetrating). Ballasted racking systems use aerodynamic measures in the design, and the weight of PV modules, racks, and any extra ballast weight calculated to be required to keep the system on the roof during wind conditions. The benefit of this type of system is that roof penetrations are minimized or eliminated, which reduces the possibility of water leaks. Also, they tend to be easier to install. Attached racking systems tend to include multiple roof penetrations. Attached systems can be an option where wind loading exceeds a ballasted system's capacity, where roof-available carrying capacity cannot accommodate a heavier ballasted system, or where an extra safety factor might be required. Another option is to include a small number of attachment points with a ballasted system to give an extra safety factor. Roof-rack systems offer the advantage of tilting PV arrays toward the equator to achieve higher annual energy production. Tilt increases energy production by allowing modules to collect more energy from the sun and by allowing snow and rain to aid in the cleaning of the module face. Typical tilts for these systems are between 5 degrees and 25 degrees.

Ground-mounted systems can be categorized as fixed or tracking. Fixed systems have lower maintenance costs but generate less energy (kWh) per unit power (kW) of capacity. Tracking systems move the PV modules so they are more directly facing the sun as the sun moves across the sky. This increases energy output but also increases maintenance and equipment costs. Single axis tracking, in which PV is rotated around a single axis, can increase energy output up to 30%. With two axis tracking, PV is able to directly face the sun all day, potentially increasing output up to 35%. Tracking is typically used only on systems of 300 kW or larger. Ground-mounted systems typically penetrate the earth, but there are applications such as ballasted tub that do not penetrate the earth and have been used where the earth is subject to settling, such as landfills. Ground-mounted systems usually allow for larger systems and greater design flexibility.

Canopy structures are frequently designed to both host PV modules and provide shade to vehicles. While this structure does provide dual utility, it will likely increase overall system cost due to the cost of the support structure. But in areas where shaded parking is valued and may be built anyway, this can be a good option.

Solar Heating and Cooling

Solar heating and cooling (SHC) technologies vary from the other solar technologies in two significant ways: 1) they produce thermal, not electric energy, and 2) the systems are not utility-scale, but are put at the customer site as distributed energy only. In buildings, SHC provides energy for water heating, space heating, and space cooling. Industrial applications include process heating or cooling for manufacturing, food processing, and other applications; district heating and cooling; and waste remediation/destruction.

This section focuses on the most common end- use applications for federal facilities and military housing. Cost-effectiveness and success depend on a number of variables, particularly the cost of competing fuels, the first cost and payback for the SHC system, incentives, and the presence of market barriers such as unfavorable zoning or building codes. Solar water heating provides the best illustration of these factors, which also apply to a greater or lesser extent to other SHC technologies. Solar thermal sales for space heating are quite small compared to the potential market, but proven products such as solar ventilation pre-heaters are available.

Types of Solar Heating and Cooling Technology

Solar thermal systems differ by the type of collector used to gather and store solar energy. There are three basic types of liquid collector systems: flat plate, evacuated, and concentrating. Air collectors use air as the working fluid for absorbing and transferring solar energy.

Glazed and Unglazed Flat Plate Collectors

Flat plate collectors are the simplest and most common type of collector. They are designed to heat water or another fluid to medium temperatures (approximately 140°F). This collector technology (Figure 1) generally consists of a thin flat-plate absorber sheet that intercepts and absorbs solar energy. The absorber may have a selective coating or it may be chemically coated to increase its solar absorption capacity. Copper pipes or tubes wind back and forth across the absorber and carry fluid through the flat plate collector. Often the pipes are painted black and bonded to the material of the flat plate collector to maximize heat absorption. The collector is covered with glass, or "glazing," that allows solar energy to pass through but reduces heat loss from the absorber. As heat builds up in the collector, it heats the fluid passing through the pipes. "Unglazed" flat plate collectors without the glass covering are best suited for low temperature applications, such as heating swimming pools. While these unglazed collectors capture a larger portion of the sun's energy, they lose a large portion of the absorbed heat because they are not insulated with a covering. These types of collectors are substantially less expensive than glazed systems.

Procuring Solar Energy: A Guide for Federal Facility Decision Makers 201

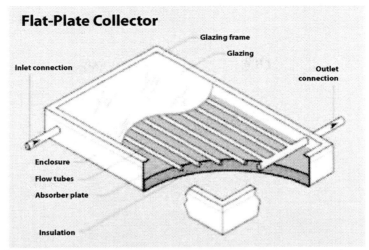

Source: DOE/EERE

Figure 1. Flat-plate collector.

Evacuated Tube Collectors

Evacuated tube collectors produce higher temperatures (approximately 300°F). This collector is made of parallel rows of tempered glass vacuum tubes and an absorber surface inside the tube. The absorber is surrounded by a vacuum that reduces heat losses. The glass tubes heat up the solar absorbers and, ultimately, the solar working fluid in order to heat domestic hot water, or provide space heating. Figure 2 is a diagram of an evacuated tube collector.

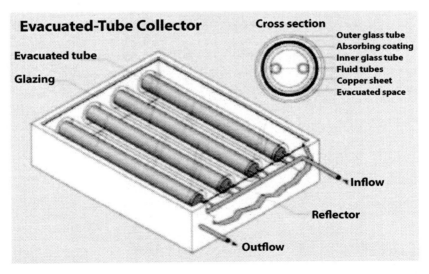

Source: DOE/EERE.

Figure 2. Flooded evacuated tube collector.

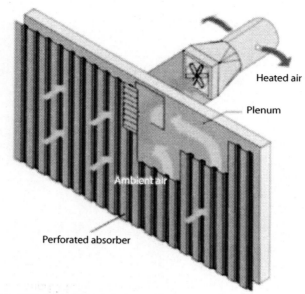

Source: Transpired Air Collectors: Ventilation Preheating, NREL, 2006.

Figure 3. Transpired air collector.

Air Collectors

Collectors for air heating systems perform the same important function as those for liquid heating systems, although they operate on a much different principle. Figure 3 shows a simple diagram of a transpired air collector. It consists of a dark-colored, perforated façade installed on a building's south-facing wall. These systems operate by using a fan or the building's own ventilation system to draw ventilation air into the building through a perforated absorber plate on the façade and up through the air space between the absorber and the south wall.

Transpired air collectors can preheat intake air by as much as 40°F. Solar roof ventilation systems are also available for both sloped and flat-roof buildings.

Space Heating

Solar heated air can be used for space heating, either directly or in combination with an air-source heat pump. One application for ventilation air is the transpired solar collector, which draws air through the perforations of a solar absorber, warming the air in the process. This heated air is then collected and used as pre-warmed air to be inputted to a conventional heating/ventilation system. They are well suited for buildings with high ventilation requirements like workshops, hangars, or warehouses. Because of its low cost and simplicity, solar ventilation air preheating using transpired collectors is cost effective in most locations that can use the heat.

Common System Components/ Configurations

For all SHC applications, the primary components are collectors, storage, and balance of systems components. Collectors are such an important component that systems are commonly characterized by the type of collector, as discussed earlier in this section. Other important components include storage, pumps, piping, heat exchangers, and mounting systems.

Water is the most common medium used for heat storage; however advancements in phase change materials could result in a shift away from water, as these synthetics have the potential to store more heat in smaller vessels. Storage tanks utilize either lined/coated steel or stainless steel in their construction. The technology utilized for solar thermal storage is much the same as traditional water heating vessels, although commercial and industrial systems may require significantly larger tanks than domestic systems. The larger the hot or chilled water storage tank, the longer the thermal supply can be considered a firm source of energy. Research is being conducted on chemical means to store heat, such as dessicant/water combinations, so that stored solar energy would be available upon request.

APPENDIX B: SELF-GUIDED SOLAR SCREENING

A preliminary solar energy site screening provides a rough estimate of the solar resource, energy production, and cost of a PV system. It also provides information for a go- or no-go decision to proceed further in the procurement process. A more thorough solar energy site screening may be needed later to better quantify the energy production and costs before proceeding further.

A summary sheet is included at the end of this section to record the data and observations. The summary sheet, along with any drawings or photographs, is needed for the next steps in procuring a PV system. Shooting photographs of the site and equipment is critical, as they allow others to confirm the preliminary estimate or make recommendations without visiting the site.

General Services Administration, Ralph H. Metcalfe Federal Building PV arrays, Chicago, Illinois. (*Credit: Patrick Engineering. PIX 09514*)

Solar Site Screening

Prior to a Site Visit

Preview the site using IMBY (www.nrel.gov/eis/imby/) or Google Earth (http://earth.google) to identify possible land or roof areas for a PV system.

- Identify roof areas with flat or equator-facing surfaces (e.g., south in the northern hemisphere) with little or no equipment on the roof.
- Identify large, open land areas.
- Print an overhead map of the site and mark these potential land and roof areas on the map for ease of location during site visit.

Use PVWATTS version 1 or 2 (www.nrel.gov/rredc/pvwatts/) or IMBY. Calculate, and print out, the energy production for a 1 kW PV system tilted at 10 degrees, and use the defaults for all other inputs. The monthly and yearly energy outputs for a 1 kW system are useful numbers for scaling to larger systems. For example, a 55 kW PV system produces 55 times the energy of a 1 kW PV system.

When on-Site

Access the roof or land area being considered for PV systems. Note the tilt angle and orientation of the equator-facing or flat roof area. Also note the type, condition, and age of the roof. If it is a land area, note the approximate grade and orientation of the land area. Take photographs.

The objective now is to determine the area of the site for a potential solar system as this will allow an estimation of the potential system size. The site needs to be very clear of objects that could cast shadows on the proposed site. While standing on the proposed site, use your thumb and fist to estimate the angle of the object from the horizon to the top of the object. A sideways thumb held at arm's length is about 2 degrees from the bottom to the top of the thumb. A fist held in front of your body is about 10 degrees from the bottom (little finger) to the top (pointer finger) of the fist (see www.vendian.org/mncharity/dir3/bodyruler_angle/). Objects that are less than 20 degrees in height above the site, that are skinny (e.g., power poles), or that can be removed should be ignored for this preliminary estimate. For objects that are on the site, make height-angle measurements close to the roof or ground where the collectors will be placed. Determine the square footage of the site that is not shaded by objects, as determined above. For a preliminary estimate, the distances could be paced off. If available, use a measuring device such as a range finder or a rolling wheel tape measure. Take several photographs that could be used to make a panoramic photograph.

If a roof area is being considered, ask the people on site if the roof leaks, when it was last replaced or repaired, or if they have any concerns about it. Note their answers. Take photographs of the roof and the underside from inside the building if possible. Ask about the roof construction and whether any drawings are available. Make a copy or take a photo of any drawings.

Photovoltaic Specific Assessment

Identify the nearest location for housing the inverters. It is best if this location is shaded or enclosed. Small inverters (6 kW or less) can hang on a wall. Larger inverters (greater than 6 kW) are placed on the ground or floor. Note the distance from the proposed PV system location to the inverter bank. Take photographs.

Identify the nearest electrical panel and record the location and distance from the inverter bank to the electrical panel, voltage at the electric panel (V), the number of phases (1 or 3), capacity of the main breaker (amps), and the capacity of the panel (amps). Take photographs of the equipment, including the circuit breakers.

Energy Production Estimate

Estimate the size of the PV system by multiplying the proposed site area (ft^2) times 9.3 W/ft^2. This corresponds to a fairly typical 14% efficient crystalline PV module. This preliminary solar energy site assessment is for no, or relatively few, solar obstructions. If the solar obstructions become numerous or complicated, then a more detailed solar energy site assessment should be made.

Site Energy Requirements

Prior to a site visit, or while on site, determine the annual energy usage from the utility bills. Ask the site personnel if any energy efficiency changes will be made or if electrical load increases are anticipated.

Determine the annual electrical energy consumption for the building or site. Compare this number to the estimated energy production from a PV system. In most locations there is little economic sense to produce more energy than is consumed. If needed, reduce the PV system size to just meet the annual electrical energy usage.

Divide the estimated PV system production by the annual electrical energy usage. This is the percentage of annual energy supplied by the PV system.

Cost Estimate

A conservative price estimate for a fully installed PV system is $6,500 to $8,000 per kW of PV. Large PV systems (greater than 100 kW) or PV systems on sites with uncomplicated site access or conditions have been installed for less money. This price range is for a simple grid connected PV system without batteries. Systems with batteries can easily double the installed price.

Incentives

Available incentives for solar projects can be critical to the economic feasibility of a prospective project. Look up and list incentives that apply to the project. The DSIRE Web site lists most incentives available for solar projects from federal, state, local, and utility sources. (www.dsireusa.org, accessed July 19, 2010)

Go- or No-go Decision

The information compiled here will form the basis of the economics that will be used for a decision to explore the feasibility of the project further.

Summary of Preliminary Solar Energy Site Screening for Photovoltaics

Name of Location

Latitude and longitude, or ZIP code:

Assessment performed by: (include contact information)

Date of Assessment

POSSIBLE SITE ISSUES

Historic building issues?

Structural issues (if rooftop)?

Roof age and condition? Planned replacement?

Area (f^2)

Maximum PV system size (kW)

Estimated annual PV system energy production (kWh/y) (from PV WATTS or INEV)

Building or site annual energy consumption (kWh/y)

Percent solar contribution (production divided by consumption) (%)

Distance from PV system to inverter (ft)

Electrical service (voltage and # of phases)

Total installed price estimate ($)

Present price of energy ($/kWh)

Estimated annual energy savings ($/y) (estimated annual energy production multiplied by present price of energy)

List available incentives for solar projects on the site:

Comments (use another page if necessary):

Attach drawings, photographs and printouts.

Solar Screening Evaluation Checklist

This is a checklist to review the adequacy and quality of a given solar screening and to determine if a more detailed screening needs to be done. If an applicable element at right is missing from the screening, it is recommended a more detailed solar screening for the site be obtained.

The checklist at right is in two parts.

The first part is for solar screenings and pertains to financing mechanisms other than agency funded. **The second part is for solar feasibility studies** and includes recommended information *in addition to* the solar screening.

Solar Screening minimum recommended information (items 1 – 5)

1. Confirm shading analysis, available square footage and preliminary size estimate:

    - Satellite map view or use of an accurate aerial tool to analyze the potential shading impacts of neighboring buildings, rooftop protrusions, parapets, or vegetation that could block sunlight from a potential solar array and relate these impacts to the available roof or ground area.
    - Satellite or accurate aerial tool measurement of the available square footage or acres
    - Preliminary estimate of the system size.

2. Confirm annual energy production per unit of capacity:

    - **Electricity production estimates**
      _____(kWh/yr/kW installed) (available online tools: PV WATTS or IMBY). Inputs for this level of analysis are as follows (unless site conditions preclude, i.e., a steep pitch roof that does not face the equator):
        - 1 kW system size
        - 10 degrees from horizontal-tilt (if you have a pitched roof, enter actual pitch)
        - Local location or nearest location option with similar sun exposure
        - 0.77 derating factor
        - Azimuth (compass orientation—select 180 degrees or 0 degrees so array faces equator or, for a pitched roof, enter the direction that the roof faces)
    - **Solar thermal** (available online tool: RetScreen www.retscreen.net/ang/home.php)
    - **Solar ventilation preheat** (available online tool: RetScreen)
    - **Solar pool heating** (available online tool: RetScreen)

3. Confirm economic analysis:
    - **Confirm local energy rates**
      _____ ($/kWh or $/therm)
    - **Annual savings**
      _____ ($/yr/ kW installed)
    - **Any available incentives?**
      _____ ($/W or $/yr/W)
      (i.e. rebates, local Renewable Energy Credits market, other). Check DSIRE: www.dsireusa.org
    - **Any extraordinary project specific costs?**
      _____ ($/W)

Confirm that structural, electrical interconnection, and equipment location issues have been investigated and any additional costs related to these have been estimated.
- Estimated System cost
  _____ ($/W)
- Appropriate economic metric for your decision-making process.
  _____ (SIR, NPV, LCC, other)

(Solar Advisor Model is available online and does some financial analysis.)

NOTE: In general, the estimated cost should not exceed $8 per watt except in special circumstances.

4. If proposed system is rooftop:
   - **Age of roof**
     _____ (yrs)
   - **Condition of roof**
     _____ (yrs of expected remaining life)
   - **Roof warranty**
     _____ (yrs remaining)
   - **Estimated structural capacity available for solar system**
     _____ (lbs/ft^2)
   - **Estimated maximum weight of solar system**
     _____ (lbs/ft^2)

5. Confirm other considerations have been addressed:
   - **Historic building issues**
     (is the proposed system on a historic building or in a historic district?)
   - **National Environmental Policy Act (NEPA)** issues (primarily an issue for large ground-mount systems).

Solar feasibility study minimum recommended information in addition to the Solar Screening information (items 1-9)

6. Confirm recommended size
   - **Is the recommended size in assessment reasonable and is there opportunity for a larger system?**
     _____ (kW or area of collectors)
     In My Backyard (IMBY) is a Web tool that uses aerial maps and a draw feature to estimate PV system size on a site. (Note: IMBY's output for system size is reasonable for a 14% efficient PV module or 100W/m^2.) See www.nrel.gov/eis/imby/ .

7. Confirm shading analysis (recommendations for report)
   - Detailed shading analysis with solar collector exclusion areas marked on the plan view of the site adjacent shading obstructions. Exclusion areas should be indicated to the east, west, and toward the equator (if in the northern hemisphere- to the south) of any shading obstruction.

- Unless the array is installed with zero degrees tilt (horizontal), need to see some space between rows in the array layout to prevent rows of PV shading each other.
8. Confirm investigation into interconnection issues:
   - Requirements to get utility approval for interconnection (estimated costs if special equipment is required)
   - Recommended interconnection point
   - Confirmation of space for system electrical equipment
9. Confirm annual energy production for site-specific recommended system
   - **Electricity production estimates**
     _____ (kWh/yr)
     (available online tools: PV WATTS or IMBY).
     Inputs are:

     - _____ (kW) system size
     - _____ (degrees from horizontal) tilt
     - _____ Location
     - _____ Derating factor (default is 0.77)
     - _____ (degrees-compass orientation) Azimuth
   - **Solar thermal** (available online tool: RetScreen)
   - **Solar ventilation preheat** (available online tool: RetScreen)
   - **Solar pool heating** (available online tool: RetScreen)

PV Project Design Evaluation Checklist

This checklist has been created to assist you in the design phase of the system during the 25%, 50%, and final review stages

Site Layout

- _____ (degrees) tilt
- _____ (degree) azimuth
- _____ (kW or MW) system size
- Engineer's stamp on PV array mounting design for wind loading
- Confirm shading analysis has been done and site layout conforms to it.

Rooftop system

- Engineer's stamp on roof structural weight carrying capacity for solar system
- Confirm weight of system is within carrying capacity of roof (lbs/sq ft)
- What type of roof penetrations, if any, and confirm construction detail to weatherproof penetrations

- Check PV layout compliance with fire specifications. (Fire Safety Guideline for Photovoltaic System Installations: www.fpemag.com/_pdf/Fire_Safety_Guideline-PV_System_Installations.pdf)
-

Ground-mount system

- _____ (ft) Height of lower edge of collector to ground. Usually like to see a minimum of 2 feet between the lower edge of the PV modules and the ground. This can be location specific (examples):
    i. In desert areas where vegetation does not grow tall and could possibly shade the array, the array could be closer to the ground.
    ii. In areas of high snow fall the array should be higher from the ground to prevent snow building up at the lower edge of the array as it slides off, thus causing the array to be shaded.
    iii. Some key things to consider regarding location are vegetation, snow, material that may drift around the array, future development, and other possibilities for future shading problems.
    iv. Decisions around this issue can be a balance between location conditions, O&M costs for periodic removal of shading problems, and specifications for the ground under the array (e.g., weed barriers and gravel).
- Perimeter fence
    i. Confirm the fence doesn't shade the array.
    ii. It is a good idea to restrict access and keep out tumbleweeds and animals that may damage the system.

For more detailed information see http://irecusa.org/fileadmin/user_upload/ National OutreachPubs/InspectorGuidelines Version2.1.pdf_

Electrical Design

Based on one-line or three-line diagram:

- Final electrical design has engineer's stamp
- _____ (volts) Check PV string maximum/minimum voltages (extreme weather) and confirm within inverter specifications (most inverters have a "string calculator" on their site, for example: www.aesolaron.com/SolarStringCalc.aspx)
- _____ (kW) Check PV array maximum DC power to each inverter (extreme weather-cold and clear) to confirm it is within inverter specifications.
- PV-each string protected (fuse or breaker)
- _____ (% efficiency, configuration & capacity) Transformer specification (recommended minimum efficiency of 97%)
- Module grounding: confirm there is a specific grounding wire or the rack/module system is Underwriters Laboratories rated for grounding
- AC disconnects specified and location easily accessible

- DC disconnects specified and location easily accessible.

Interconnection: Inside building must meet NEC 690
(Sum of PV breaker and panel main breaker less than or equal to 120% of panel rating)

- _____ (amps) Panel rating
- _____ (amps) Panel main breaker rating
- _____ (amps) PV Breaker

Interconnection: Direct tie

- _____Is there an acceptable plan?

System Components Specifications

- System components meet "Buy American" criteria?
- PV CEC approved?
- Inverter(s) CEC approved?
- _____% Inverter efficiency
- _____ (years) Inverter warranty (recommended 10 year minimum)
- _____PV module warranty (recommended 10 year 90%, 20 year 80% minimum)

PV Commissioning Checklist

PV system commissioning should do three primary things: 1) verify the system meets contractual requirements, 2) determine whether the system meets safety requirements, and 3) confirm that the system performs as expected. The two basic components of a commissioning are visual inspection and performance testing.

Before Commissioning

The items below should be completed before commissioning or available at the site.

- All permits have been signed off
- Utility has given permission to operate system
- One or three line drawing of system
- System layout drawing (shows module layout, location of balance of system components, disconnects, and wiring and conduit specifications)
- PV module specifications
- Inverter specifications
- Combiner box specifications
- Feasibility Study Review (if available).

Visual Inspection

Verify the installation is complete to the design drawing.

- All PV modules are permanently installed (confirm modules are in good condition)
- All inverters permanently installed
- All combiner boxes permanently installed
- All disconnects and switchgear permanently installed
- Wiring is completed (no loose connections or damaged wires)
- No potential for wire damage (e.g., deburred metal and proper sheathing to protect wires)
- Utility power connected
- Internet connection operational (if applicable)
- Physical installation is per design drawing and manufacturer's specification, and it meets visual requirements
- System is compliant with applicable building and electrical codes
- Protective fencing and enclosures are installed
- Verify outdoor equipment is of proper material for location (e.g., UV-rated wire, stainless steel, hot dipped galvanized, other as specified)
- Dissimilar metal should be electrically isolated to avoid galvanic corrosion
- Verify grounding of metallic surfaces that might become energized
- Wire and conduit sizes installed per plan
- Fuses and breakers are sized and installed properly
- Document as-built conditions
- All equipment is labeled as required.

For a more detailed description of a similar procedure see http://solarprofessional.com/files/sample/sp2_6_pg34_gleason.pdf.

Performance Testing

Verify the system is performing within acceptable limits.

- Conductor insulation test using a megohm meter on all homerun wiring to ensure no leakage currents to earth (Pos-to-GND and Neg-to-Gnd resistance > 2 megaohms). This may need to be done during construction while conductors are accessible.
- Grounding resistance is < 5 Ohms
- Measure and record open-circuit voltage (V_{oc}) and polarity of each string. (Verifies all strings have the same number of modules.)
- Measure and record short-circuit current (I_{sc}) of each string.
- Inverter startup sequence – follow manufacturer's instructions for initial startup.
- Measure and record maximum power point current (I_{mp}) for each string. (Current measurements for each string should be within a 0.1A range of each other, assuming consistent weather conditions and all string having same tilt and azimuth angle. If a string is outside the range, check for shading or a ground fault.)

- Confirm inverter's internal power meter and display using independent meters. (Once this is done, inverter-displayed power readings can be used for subsequent reporting.)
- Confirm the system output under actual conditions meet minimum expected output. Actual performance should be within about 5% of expected, calculated performance. This procedure includes system nameplate rating (kW), solar irradiance measurement (W/m^2) and module cell temperature (C). Procedure is best conducted during consistent weather conditions, where no array shading is present, and solar irradiance is not less than 400 W/m^2.

APPENDIX C

Agency-Funded Project Case Studies

U.S. General Services Administration (GSA), Denver Federal Center PV Project, Denver, Colorado, 1.19-MW PV

- This PV system generated 1,726,000 kWh in 2008, 14% more than the contract required (10% of the DFC campus peak electrical load, 2.5% of total Denver Federal Center kWh)
- Collectors at 20 degree fixed tilt
- Requires 6 acres of land
- First cost was $6.9 million ($5.8\Watt), GSA owned
- Incentives - $200K
- REC = 1,525 MWh, $240/MWh for 20 years

More information is available at www.gsa.gov/portal/content/105165 (Photo Credit: Dave Mowers; U.S. General Services Administration (GSA). PIX 17421)

Social Security Administration, high- temperature hot water project, Philadelphia, Pennsylvania, 54-m² solar thermal system

- Reheats recirculation loop
- 360 evacuated heat-pipe collector tubes, 54 m² gross area, 36 m² net absorber area
- Cost $58,000
- Delivery of 143 million Btu/year estimated
- Installed 2004

More information is available at www.eere.energy.gov/femp/services/yhtp/energy_projects_detail.cfm/id=8 (Photo Credit: Ed Hancock, Mountain Energy Partnership.)

Power Purchase Agreement Case Studies

National Renewable Energy Laboratory PV Project, Golden, Colorado, 720-kW PV
- 720 kW (1200 MWh) single-axis tracking, approximately 5 acres
- 20-year power purchase agreement (PPA) contract (utilizing Western)
- 20-year easement
- RECs sold to Xcel Energy for RPS solar set-aside (20-year contract)
- PPA price equal to or less than utility electricity prices (based on EIA projections)
- Operational December 2008
- Additional PV projects in progress.

More information is available at www.eere.energy.gov/femp/pdfs/pfs_mesatoparray.pdf (Photo Credit: SunEdison, PIX 17423)

Fort Carson PV Project, Colorado Springs, Colorado, 2-MW PV

- 2 MW, 3200 MWh in first year (~2% of Ft. Carson's load)
- Fixed, non-escalating energy rate
- 17-year contract, with 3-year option (utilizing Western)
- No-cost, 20-year lease (using 10 USC 2667 lease authority)
- RECs sold to Xcel Energy (20-year contract)
- Ground-mounted, fixed system covering 12-acre former landfill
- First Solar thin film, 25-year warranty
- Came on-line December 2007

More information is available at www.3phases.com/news/news-item.php?id=32 (Photo Credit: U.S. Army Fort Carson. PIX 17394)

Energy Savings Performance Contract Case Studies

U.S. Marine Corps Base PV Project, Twenty-nine Palms

- 1.1 MW solar energy project
- Army Corps ESPC agreement awarded to Johnson Controls in 2002.
- PV project cost $8.1 million after a rebate of $4.5 million from the State of California
- Annual guarantee of $400,000 in energy cost
- Total savings from the project to date equal $6.9 million.
- The PV system utilizes 8700 modules sited on 6.5 acres.
- The system is tied into the cogeneration plant at 12.47 kV.

More information is available http://tonto.eia.doe.gov/kids/energy.cfm?page=solar_home-basics-k.cfm and www.ornl.gov/sci/femp/pdfs/fs-5903_29palms_usmc.pdf (Photo Credit: Daniel C. Kariuki, Energy Projects Manager, Marine Corps Air Ground Combat Center, 29 Palms, California)

Phoenix Federal Correctional Institution, Phoenix, Arizona, parabolic trough solar water-heating system

- 17,040 square feet of parabolic trough collectors
- 23,000-gallon storage tank
- Installed cost of $650,000
- Delivered 1,161,803 kWh in 1999 (87.1% of the water heating load)
- Saved $77,805 in 1999 utility costs

More information is available at www.eere.energy.gov/femp/pdfs/33211.pdf
(Photo Credit: Ed Hancock, Mountain Energy Partnership. PIX 09048)

Utility Energy Services Contract Case Studies

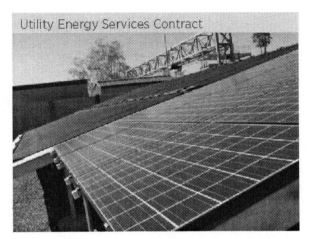

Camp Pendleton, North San Diego County, California, PV Project, 75kW PV

- 75 kW PV project with 116,000 KWh/year estimated production (actual production has been higher)
- Bundled with various EE measures
- Total project cost was $11.2 million,
- Simple payback 7.8 years, saving investment ratio (SIR) 1.94
- Contract term is 10 years
- Projected annual savings is 62,377 MMBtus.
- Total projected California solar incentive covers approximately 33% of project costs
- Online since July 2008

More information is available at www.eere.energy.gov/femp/pdfs/46348.pdf
(Photo Credit: U.S. Marine Corps. PIX 16462)

Joshua Tree National Park, California, PV/propane hybrid

- 20.5 kW PV array
- 613 kWh battery bank
- 35 kW propane generator
- $273,000 cost financed by Southern California Edison under 15-year tariff

More information is available at www.eere.energy.gov/femp/pdfs/46348.pdf
(Photo Credit: Harry Carpenter. PIX 07260)

Enhanced Use Lease Case Studies

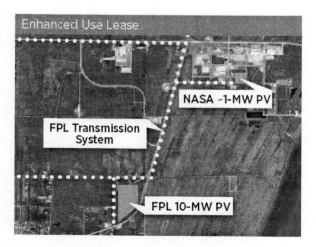

NASA Kennedy Space Center/Florida Power & Light Company (FPL), Merritt Island, Florida, 10-MW PV, owned by FPL with in-kind consideration of 990-kW NASA-owned PV

- Partnership between NASA Kennedy Space Center and Florida Power & Light

- EUL signed June 2008
- Phase 1 involves 60 acres; potential phase 2 for additional 40 acres
- 10 MW FPL-owned PV project
    - Construction started; estimated completion date 3/10
    - Output feeds into FPL transmission system
    - Substation expansion required

- In-kind consideration - 990 kW NASA-owned PV – FPL construction, O&M
    - Construction started, estimated completion date 10/09
    - Output feeds into NASA
    - owned distribution system

- 130 mph wind standard

More information is available at www.smartgridnews.com/artman/uploads/1/nasa_space_coast_solar.pdf and www.fpl.com/environment/solar/spacecoast.shtml (Photo Credit: NASA Kennedy Space Center/Florida Power & Light.)

Fort Irwin (in process), Barstow, California, 500-MW solar thermal/PV by 2022

- Notice of opportunity to lease
- Approximately 500 MW at five Fort Irwin sites
- In-kind services equal to or greater than fair market value of land
- Developer conducts NEPA EIS
- Fort Irwin will conduct an environmental baseline study as a part of the lease documents
- Developer selection announced July 30, 2009 – Clark and Acciona Solar Power
- First phase:
  More than 500 MW solar thermal/PV by 2022

More information is available at http://eul.army.mil/ftirwin/ (Photo Credit: Google maps)

CHAPTER SOURCES

The following chapters have been previously published:

Chapter 1 – This is an edited, excerpted and augmented edition of a United States Congressional Research Service publication, Report Order Code R41297, dated October 1, 2010.

Chapter 2 – This is an edited, excerpted and augmented edition of a United States Congressional Research Service publication, Report Order Code R41197, dated April 20, 2010.

Chapter 3 – This is an edited, excerpted and augmented edition of a United States Government Accountability Office publication, Report Order Code GAO-10-196T, dated October 27, 2009.

Chapter 4 – These remarks were delivered as Statement of Casey Coleman, before the U.S. House of Representatives Committee on Oversight and Government Reform, dated October 27, 2009.

Chapter 5 – These remarks were delivered as Statement of James Jones, before the U.S. House of Representatives Committee on Oversight and Government Reform, dated October 27, 2009.

Chapter 6 – These remarks were delivered as Testimony of Rick Goss, before the Military Personnel Subcommittee, House Armed Services Committee, U.S. House of Representatives, dated July 23, 2008.

Chapter 7 – These remarks were delivered as Statement of Jeff Omelchuck, before the Military Personnel Subcommittee, House Armed Services Committee, U.S. House of Representatives, dated July 23, 2008.

Chapter 8 – This is an edited, excerpted and augmented edition of a United States Department of Energy publication, dated September 2010.

INDEX

A

abatement, 101
absorption, 200
access, 38, 50, 52, 131, 146, 151, 155, 157, 159, 171, 179, 191, 193, 199, 205, 210
accountability, 87, 91, 92, 111
accounting, 35, 56, 58, 124, 134, 135, 189
accuracy, 48, 65
acid, 70, 77, 83
acquisitions, 2, 3, 18, 24, 27, 29, 30, 37, 38, 39, 49, 51, 57, 58, 59, 62, 64, 70, 73, 111
adaptation, 35
adhesives, 79
adjustment, 22
advancements, 203
advantages, 35, 40, 101
adverse effects, 137
aesthetic, 134
aesthetics, 199
affirmative action, 1, 3, 21
age, 104, 130, 204
Agency for International Development, 56
aggregation, 33, 185, 193
air emissions, 54, 55, 113
Air Force, 25, 26, 124
air pollutants, 43
alcohols, 5
American National Standards Institute (ANSI), 46
American Recovery and Reinvestment Act, 27
American Recovery and Reinvestment Act of 2009, 27
American Samoa, 24
animal welfare, 68
appropriations, 127, 135, 191
ARC, 72
asbestos, 41, 68, 72
Asia, 77
Asian countries, 83

assessment, vii, 29, 37, 40, 41, 42, 43, 44, 45, 47, 48, 61, 62, 66, 69, 76, 78, 118, 119, 126, 133, 165, 166, 167, 182, 194, 205, 208
assets, viii, 72, 73, 82, 87, 88, 91, 96, 97, 165, 188
atmosphere, 17, 66, 184, 185, 187
attachment, 154, 199
audit, 60, 77, 79, 81, 84, 96, 119, 137, 165, 167, 171, 173, 187
audits, 173
authorities, vii, 1, 2, 4, 5, 7, 8, 9, 12, 13, 15, 22, 25, 30, 64, 176
authority, 8, 9, 12, 13, 20, 22, 23, 26, 62, 95, 129, 132, 137, 150, 151, 152, 153, 156, 176, 177, 183, 186, 215
automobiles, 88
avoidance, 21, 196
awareness, 47, 48, 100, 109

B

background information, 196
banking, 169
barriers, 31, 48, 49, 70, 92, 93, 104, 122, 129, 200, 210
base, 27, 99, 104, 105, 124, 167, 189
baths, 77, 83
batteries, 108, 186, 197, 205
benchmarks, 93
benefits, 5, 43, 46, 48, 70, 76, 78, 79, 80, 81, 82, 83, 84, 85, 87, 88, 93, 94, 98, 100, 104, 109, 113, 125, 150, 158, 168, 172, 174, 188, 191, 192, 194
biological systems, 41
biomass, 122, 185, 192
bonds, 185
brand development, 114
brand image, 114
Brazil, 94, 110
building code, 35, 194, 200
buildings, 3, 30, 31, 32, 34, 40, 41, 43, 57, 65, 67, 68, 70, 93, 123, 125, 127, 137, 143, 184, 185, 187, 189, 193, 196, 199, 200, 202, 207

Bureau of Land Management, 118
burn, 22
businesses, 1, 3, 4, 22, 32, 65, 69, 92, 100, 184, 185, 186, 190, 194, 195
buyer, 3, 13, 35, 38
buyers, 4, 14, 35, 38, 78, 120, 121, 137, 196
by-products, 6

C

cadmium, 77
candidates, 37, 193
capital projects, 140
carbon, 6, 99, 101, 102, 103, 123, 124, 134, 156, 185, 188, 190
carbon dioxide, 99, 123, 134, 188, 190
carbon emissions, 101, 102, 103
carbon tetrachloride, 6
case studies, 48, 121, 139
cash, 187
category a, 55
CEC, 118, 142, 211
cell phones, 77, 94, 110, 114
certification, vii, 17, 23, 27, 29, 44, 45, 46, 51, 64, 69, 70, 78, 125, 127, 137, 143
challenges, 8, 103, 105, 120, 129
Chamber of Commerce, 25
chemical, 6, 17, 27, 41, 42, 54, 66, 67, 203
chemicals, 6, 10, 16, 17, 23, 30, 32, 37, 38, 45, 51, 57, 67, 94
Chicago, 203
child labor, 3
children, 84
chimneys, 194
China, 83, 94
cities, 108, 193
citizenship, 69
City, 66, 68, 69, 70
civilization, 41
class, 22, 54
classification, 8, 25
Clean Air Act, 4, 20, 71
clean energy, viii, 73, 103, 120, 125
Clean Water Acts, 2, 12, 13
cleaning, 37, 42, 66, 199
clients, 108
climate, 41, 98, 125, 187, 189, 193
climate change, 125, 187
CO2, 101, 185, 187, 190
coal, 5, 33, 43
Coast Guard, 130
cogeneration, 167, 216
combustion, 16, 185

commerce, 83, 102
commercial, 2, 13, 15, 21, 23, 27, 37, 38, 72, 93, 99, 102, 114, 143, 148, 150, 185, 203
communities, 103, 109
community, 17, 67, 88, 92
competing interests, 3, 21, 70
competition, 3, 4, 8, 17, 19, 21, 22, 25, 97, 98, 193
competitive advantage, 142
competitiveness, 103
competitors, 153
compilation, 26
complexity, 13, 44, 45
compliance, 26, 27, 59, 60, 72, 99, 108, 109, 111, 112, 115, 132, 182, 184, 185, 210
compounds, 185
computer, 25, 27, 38, 57, 58, 78, 79, 80, 81, 83, 92, 93, 94, 95, 110, 111, 112, 137
computer systems, 38
computing, 58, 89, 98, 113
conceptual model, 43, 45
conditioning, 39, 104
conductors, 212
configuration, 159, 194, 210
configurations, 105
Congress, 1, 2, 7, 12, 21, 22, 29, 31, 44, 59, 60, 61, 62, 63, 64, 65, 68, 72, 73, 100, 101, 104, 107, 152
connectivity, 132
consensus, 34, 46, 66, 92, 99, 108, 114, 189
conservation, 27, 39, 55, 57, 73, 78, 118, 164, 165, 185, 187
consolidation, 87, 88, 105
constituents, 52, 53
Constitution, 2, 8, 26
construction, 3, 7, 13, 21, 24, 27, 33, 58, 65, 67, 68, 73, 137, 149, 150, 155, 156, 157, 159, 163, 168, 169, 171, 175, 182, 194, 199, 203, 204, 209, 212, 219
consulting, 27
consumer electronics, 114
consumers, 31, 48, 50, 70, 92, 93, 94, 100, 102, 103, 107, 109, 132, 186, 190
consumption, 16, 41, 58, 65, 79, 89, 100, 105, 123, 132, 154, 160, 195, 205
containers, 70, 80
Continental, 21
contingency, 27
controversial, 41, 68
controversies, 31, 61
cooling, 32, 67, 104, 119, 143, 148, 194, 196, 200
coordination, 102, 154, 163, 175, 181, 182
copper, 77, 83
corrosion, 212

cost saving, 18, 24, 61, 76, 80, 87, 100, 101, 127, 167
cotton, 25
counsel, 21, 129, 158
covering, viii, 14, 22, 53, 110, 200, 215
creativity, 129
crops, 68
crystalline, 197, 198, 205
current limit, 64
customers, 14, 27, 44, 102, 105, 188, 189, 191, 194, 195
cycles, 40, 43, 47, 62, 195
cycling, 115

D

damages, 119, 154, 158, 190
data center, 89, 104, 105
database, 33, 50, 52, 53, 55, 56, 59, 69, 80, 84, 110
decentralization, 48
decision makers, 120, 121, 126
decision-making process, 128, 208
defendants, 27
deficiencies, 60, 184
degradation, 157, 159
demonstrations, 125
Department of Agriculture, 6, 13, 24, 38, 53, 55, 67, 71, 72
Department of Commerce, 112
Department of Defense, 20, 21, 56, 71, 72, 118, 151, 172, 174
Department of Energy, 6, 31, 39, 53, 54, 65, 67, 68, 69, 71, 72, 76, 78, 92, 105, 112, 117, 118, 121, 122, 193, 195, 196, 221
Department of Labor, 112
Department of State, 112
Department of the Interior, 112
Department of Veterans Affairs, 27, 111, 177
Departments of Agriculture, 80
depreciation, 151, 158
depth, 76, 82, 83
destination, 20, 51, 58, 59, 71, 73, 78, 80, 96
destruction, 43, 200
developing countries, 77, 78
diesel fuel, 5
disadvantages, 64
disaster, 27
disaster relief, 27
discharges, 95
disclosure, 172
discrimination, 3, 9
disposition, 82, 85, 95, 96, 158

distribution, 21, 36, 100, 101, 132, 186, 191, 193, 195, 197, 219
district heating, 200
District of Columbia, 24
DOC, 112
DOI, 112
DOL, 112
dominance, 58
Dow Jones Sustainability Index, viii, 97
draft, 7, 9
drawing, 211, 212
due process, 2, 8, 26

E

ECM, 118, 169, 187
e-commerce, 83
economic competitiveness, 103
economic incentives, 105
economic theory, 188
economics, 135, 168, 172, 173, 174, 206
economies of scale, 46, 126, 152
economy, viii, 30, 31, 49, 61, 73, 98, 103, 120, 125
ecosystem, 41, 55, 66, 68
education, 48
educational institutions, 78
Efficiency, 32, 39, 57, 66, 68, 70, 104, 118, 124
efficiency criteria, 115
efficiency level, 54
electric charge, 184, 186
electricity, 5, 43, 48, 57, 67, 88, 89, 94, 103, 104, 113, 118, 120, 121, 122, 123, 124, 131, 132, 133, 134, 136, 141, 150, 151, 152, 155, 156, 165, 184, 185, 186, 187, 188, 189, 190, 191, 192, 193, 195, 196, 198, 215
electronic devices, vii, 75
Electronic Product Environmental Assessment Tool (EPEAT), viii, 18, 24, 57, 87
emergency, 196, 197
emission, 20, 54, 123, 124
employees, 3, 76, 80, 81, 82, 88, 100, 112, 114, 143
EMS, 59, 60, 62, 71
endangered species, 68, 188
end-users, 47
energy consumption, 65, 89, 100, 105, 123, 205
energy efficiency, vii, viii, 2, 3, 6, 24, 29, 43, 45, 65, 71, 73, 78, 92, 93, 97, 98, 99, 100, 103, 104, 105, 108, 109, 110, 113, 115, 118, 121, 122, 123, 135, 136, 152, 164, 166, 170, 171, 173, 187, 191, 194, 205
Energy Independence and Security Act, 20, 24, 32, 34, 71, 118, 123
energy management system, 167

energy markets, 135
Energy Policy Act of 2005, 24, 32, 71, 118, 122
energy prices, 121, 150, 151
energy savings performance contract (ESPC), 126, 187
Energy Star, vii, 6, 10, 14, 18, 20, 22, 24, 28, 29, 30, 31, 32, 36, 39, 45, 46, 48, 49, 51, 53, 54, 55, 57, 58, 63, 65, 66, 69, 71, 73, 76, 78, 83, 84, 91, 92, 95, 98, 99, 100
enforcement, 7, 19, 187
engineering, 14, 130, 142, 181
environment, vii, viii, 1, 2, 6, 15, 16, 17, 26, 29, 32, 34, 37, 39, 42, 43, 50, 54, 55, 60, 65, 66, 67, 68, 69, 71, 72, 73, 75, 77, 78, 84, 88, 92, 97, 102, 103, 109, 120, 121, 125, 177, 190, 219
environmental advocates, viii, 109
environmental aspects, 59
environmental effects, 37, 39
environmental factors, vii, 1, 2, 9, 160
environmental footprints, vii, 29, 35
environmental impact, vii, viii, 7, 30, 34, 35, 37, 38, 39, 40, 42, 43, 47, 52, 59, 75, 77, 81, 91, 92, 107, 109, 118, 119, 122, 134, 182, 190
environmental issues, 35, 109, 145, 155
environmental management, 9, 11, 59, 62
environmental performance, viii, ix, 46, 50, 60, 78, 79, 83, 93, 97, 110
environmental policy, 35
environmental protection, 20, 34
Environmental Protection Agency, viii, 6, 12, 20, 22, 23, 30, 31, 38, 51, 53, 54, 55, 56, 66, 67, 68, 69, 70, 71, 72, 73, 75, 77, 80, 87, 91, 112, 118, 124
environmental regulations, 34
environmental standards, 115
EPA, viii, ix, 6, 8, 12, 13, 14, 18, 19, 22, 23, 24, 26, 30, 31, 32, 38, 39, 40, 41, 45, 50, 51, 52, 53, 54, 55, 57, 60, 66, 67, 68, 69, 71, 72, 73, 75, 76, 77, 78, 79, 80, 81, 82, 83, 84, 85, 91, 92, 93, 94, 95, 96, 99, 100, 101, 102, 104, 107, 108, 109, 110, 111, 112, 113, 114, 115, 118, 124
EPEAT Bronze, viii, 93, 110, 111
EPEAT Gold, viii, 99, 110, 111
EPEAT Silver, viii, 110, 111, 112
EPEAT Standard, viii, 110
equipment, 15, 27, 30, 31, 33, 40, 41, 48, 53, 58, 66, 67, 76, 77, 78, 79, 80, 81, 82, 83, 84, 92, 94, 95, 96, 98, 99, 101, 102, 104, 112, 123, 131, 132, 134, 137, 143, 145, 150, 156, 158, 173, 184, 186, 187, 193, 194, 197, 199, 203, 204, 205, 208, 209, 212
erosion, 42, 68
ethanol, 5
EU, 115

Europe, 94
European Commission, 67, 70
European Union, 55
evacuated tube collectors, 143
exclusion, 133, 208
executive branch, viii, 7, 20, 61, 62, 79, 120
Executive Office of the President, 80
Executive Order, 6, 11, 14, 20, 21, 22, 23, 24, 26, 51, 52, 58, 59, 63, 65, 67, 71, 72, 76, 82, 85, 91, 99, 100, 101, 104, 107, 108, 110, 111, 122
executive orders, 4, 8, 13, 22, 24, 29, 31, 50, 51, 52, 59, 61, 62, 65, 71, 110, 120
exercise, 8, 25
expenditures, 30, 51, 56, 58, 188
expertise, 13, 93, 117, 173
exports, 80
exposure, 207
externalities, 42, 49
extraction, 68, 83

F

fabrication, 68
facilitators, 173
fairness, 21, 166
Federal Acquisition Regulation (FAR), 3, 5, 20, 107
federal agency, 71, 75, 77, 80, 82, 84, 85, 104, 125, 152, 170
Federal agency leaders, viii, 120
federal electronics challenge (FEC), 76, 78
federal facilities, 2, 76, 80, 81, 95, 121, 200
federal government, 2, 4, 15, 20, 21, 25, 27, 29, 30, 31, 38, 45, 46, 50, 51, 57, 60, 61, 63, 64, 69, 73, 75, 76, 78, 83, 84, 97, 98, 99, 101, 102, 108, 122, 123, 125, 135, 137
Federal Government, viii, 20, 21, 25, 72, 84, 91, 94, 95, 96, 97, 99, 111, 114, 123
Federal Government's purchasing practices, viii, 91
federal law, 35, 47, 150
federal procurement, vii, 1, 2, 3, 4, 5, 12, 20, 21, 22, 29, 30, 31, 49, 51, 56, 61, 63, 75, 76, 78, 101, 166
federal programs, 30, 32, 61
Federal Register, 51, 59, 71
federal regulations, 31
federal spending, vii, 1, 2
federal workforce, 76, 82
feedback, 117, 153, 192
FEM, 57
financial, 49, 93, 101, 114, 136, 141, 146, 154, 157, 160, 171, 178, 179, 180, 190, 208
financial incentives, 136
financial support, 93, 101
Financial Times Sustainability Index, viii, 97

fisheries, 42
fishing, 37, 67
flaws, 60, 148, 163, 168, 175, 182
flexibility, 61, 62, 79, 199
fluid, 42, 200, 201
food, 6, 41, 54, 55, 72, 200
force, 20, 72
formaldehyde, 40
formula, 22
framing, 43
France, 68
freshwater, 40, 41
fuel cell, 22
fuel consumption, 16, 58
fuel economy, 39
funding, 45, 82, 100, 101, 108, 109, 115, 117, 133, 138, 140, 141, 152, 169, 190, 191
funds, 21, 30, 82, 121, 140, 151, 153, 162, 192

G

GAO, 9, 70, 73, 75, 76, 84, 85, 221
General Services Administration (GSA), 30, 38, 126, 134, 213
geography, 25
Global 100, viii, 97
global climate change, 187
global leaders, 97
global warming, 30, 185
goods and services, 1, 3, 8, 14, 15, 29, 30, 51, 69
google, 204
Google, 204, 219
government policy, 73
government procurement, 21
government staffers, viii, 109
governments, 2, 21, 30, 45, 47, 64, 70, 99, 181
grants, 93, 135, 186, 193
Green building, vii, 29
green buildings, 60
green procurement, vii, 22, 29, 30, 31, 32, 34, 35, 36, 37, 41, 42, 43, 44, 45, 46, 47, 48, 49, 50, 51, 52, 58, 59, 60, 61, 62, 63, 64, 65, 66, 67, 68, 70, 71, 73, 91
Green Rankings of America's 500 largest corporations, viii, 97
greenhouse, 3, 15, 16, 25, 31, 66, 67, 70, 83, 88, 92, 93, 94, 95, 113, 118, 123, 185, 187
greenhouse gas emissions, 3, 16, 25, 83, 92, 93, 94, 113, 123
greenhouse gases, 66, 70, 88, 95, 187
greening, viii, 57, 58, 85, 91, 96, 110
grounding, 210, 212
growth, 30, 114

GSA, viii, 14, 15, 26, 27, 28, 30, 38, 52, 53, 54, 55, 59, 60, 72, 73, 78, 87, 88, 89, 112, 117, 118, 126, 134, 142, 143, 144, 146, 171, 172, 213
guidance, 8, 13, 18, 23, 26, 30, 48, 50, 51, 52, 59, 63, 66, 70, 79, 84, 121, 123, 124, 126, 134, 135, 148, 155, 164, 176
guidelines, 19, 26, 37, 48, 54, 66, 82, 139, 148
guiding principles, 68

H

handheld devices, 77
harmonization, 64
harmony, 65
harvesting, 67
hazardous materials, 79, 83
hazardous waste, 41, 78, 81, 94, 113
hazardous wastes, 41
hazards, 41
headquarters, 82, 112, 122, 153
health, 17, 30, 34, 37, 39, 40, 41, 69, 77, 103, 105, 125, 157, 179
Health and Human Services, 56, 80
health condition, 77
health effects, 37
health problems, 84
health risks, 40, 41
heat loss, 200, 201
heavy metals, 41
height, 131, 143, 204
hemisphere, 204, 208
history, 21, 109, 164, 170, 176, 180
homeland security, 27
homeowners, 92
homes, 32, 93, 94, 113, 186, 194, 195
Hong Kong, 78
host, 126, 199
hotels, 93
House, 20, 24, 67, 68, 72, 85, 91, 96, 221
House of Representatives, 221
housing, 32, 93, 200, 205
hub, 108
human, viii, 6, 16, 29, 34, 37, 41, 42, 50, 55, 66, 67, 75, 77, 78, 84, 89, 102, 140
human activity, 34
human health, viii, 6, 16, 29, 37, 42, 50, 55, 66, 67, 75, 77, 78, 84, 89, 102
hurricanes, 198
hybrid, 16, 22, 172, 218
hydrogen, 5

I

identification, 13, 25, 120, 130, 153
identity, 22, 114
IEEE 1680, ix, 94, 99, 110, 114
image, 114
impact assessment, 43
impacts, vii, 29, 34, 35, 37, 38, 40, 41, 42, 43, 47, 48, 67, 68, 69, 70, 76, 95, 98, 100, 107, 108, 109, 182, 207
improvements, 37, 98, 103, 105, 120, 121, 150, 166, 168, 170, 195
income, 152, 158, 185, 187, 188
income tax, 185, 188
independence, 30
Independence, 20, 24, 32, 34, 71, 118, 123
India, 78, 83
indirect costs, 42
indirect effect, 41
individuals, 185
industries, 30, 35, 61, 63, 85, 98, 109
industry, viii, 21, 23, 33, 34, 41, 45, 49, 50, 58, 66, 71, 79, 88, 101, 103, 105, 109, 120, 121, 131, 150, 153, 178, 179, 198
industry/manufacturers, viii
inflation, 162
information technology, 30, 35, 56, 57, 58, 64, 77, 87, 92, 94, 97, 103, 112
infrastructure, 87, 102, 103, 104, 144, 176, 188, 194
Institute of Electrical and Electronic Engineers, ix, 55, 57, 110
institutions, 38, 46, 47, 99, 194
insulation, 149, 163, 169, 175, 183, 212
integration, 36, 42, 43, 64, 198
integrity, 2, 8, 21
intelligence, 7, 19
internal rate of return, 119, 131
International Organization for Standardization (ISO), 46
international standards, 59
Internet, 212
inversion, 198
investment, 70, 71, 103, 107, 119, 125, 126, 127, 136, 141, 151, 157, 158, 165, 171, 188, 189, 217
investment ratio, 119, 174, 217
investments, 73, 99, 188
investors, 154
Iran, 21
issues, vii, 31, 35, 60, 63, 71, 105, 109, 115, 128, 129, 130, 131, 132, 137, 145, 148, 149, 151, 152, 155, 157, 162, 163, 169, 171, 175, 179, 180, 182, 183, 188, 189, 198, 208, 209
IT procurement, viii, 57, 58, 91, 98

J

Japan, 55, 94
job performance, 104
jurisdiction, 184
justification, 17, 19, 119, 172

L

labeling, 29, 31, 36, 39, 40, 44, 45, 46, 47, 50, 64, 194
landfills, 38, 77, 80, 84, 88, 89, 199
landscape, 66
laptop, 57, 76, 78, 79, 80, 83, 89
law enforcement, 7, 19
laws, 2, 14, 35, 78, 89, 150, 187
laws and regulations, 150
LD, 119, 158, 190
lead, vii, viii, 3, 40, 42, 44, 52, 55, 57, 75, 77, 83, 84, 85, 92, 97, 101, 120, 125, 142, 145, 151, 155
leadership, viii, 91, 99, 122, 125, 129
leadership characteristics, 129
leakage, 43, 212
leaks, 199, 204
learning, 84, 102, 151
LEED, 119, 125, 127, 189
legal issues, 152
legality, 150
legislation, 2, 61, 120, 150, 172, 174
life cycle, vii, 12, 29, 36, 37, 38, 39, 40, 41, 42, 43, 44, 45, 46, 47, 50, 55, 57, 61, 62, 63, 64, 65, 66, 69, 87, 123, 131
lifecycle approach, viii, 87
lifetime, 104
light, 15, 79, 195, 196
Limitations, 15, 16, 17, 18, 19
liquid fuels, 5
Lithuania, 110
local government, 2, 30, 47, 64, 72, 99, 194
logging, 37
longevity, 108, 197

M

magazines, 101
magnitude, vii, 1, 2
majority, 80, 198
management, viii, 3, 9, 11, 21, 26, 27, 32, 37, 38, 55, 57, 58, 59, 62, 63, 64, 69, 70, 75, 76, 77, 78, 80, 81, 82, 87, 88, 91, 93, 95, 97, 98, 100, 102, 109, 114, 119, 122, 125, 135, 167, 177, 181, 182, 187
manufacture, viii, 37, 66, 91, 99, 102, 109

manufactured goods, 30
manufacturing, 1, 3, 6, 16, 25, 36, 37, 41, 46, 68, 92, 107, 108, 109, 186, 200
Marine Corps, 119, 137, 151, 165, 167, 216, 217
marketing, 14, 38, 50, 70, 108, 114
marketplace, 30, 37, 42, 44, 49, 63, 71, 92, 93, 98, 101, 102
Maryland, 197
MAS, 27
materials, 2, 3, 5, 6, 23, 25, 26, 27, 32, 36, 37, 38, 39, 40, 41, 47, 51, 52, 54, 55, 57, 61, 63, 67, 68, 69, 70, 71, 72, 73, 76, 78, 79, 83, 92, 94, 95, 98, 102, 109, 113, 114, 137, 184, 185, 203
matrix, 43, 44
matter, 20, 25, 40, 73, 82, 122, 185
measurement, 32, 93, 119, 167, 190, 207, 213
measurements, 142, 204, 212
medical, 42
Members of Congress, vii, 1, 2
memorandums of understanding, 111
memory, 79
mercury, 43, 57, 77, 79, 92, 94, 113
metabolized, 42
metals, 41, 77, 83, 94, 102
meter, 150, 153, 184, 190, 194, 195, 212, 213
methanol, 5
methodology, 40, 105
Mexico, 110
military, 25, 27, 167, 200
mining, 37, 41, 67, 92
mission, 87, 163, 169, 175, 180, 183
missions, 3, 23, 73, 123, 134, 187, 190
misunderstanding, 146, 159
mobile device, 57
modeling, 69, 92
models, 44, 88, 105
modernization, 87, 104
modifications, 62
modules, 143, 158, 159, 167, 184, 190, 196, 198, 199, 210, 212, 216
monitoring, 23, 59, 60, 137, 150, 164, 170, 176, 183
morale, 88
motivation, 50, 169
multidimensional, 36
multiplier, 185
municipal solid waste, 122

N

NAFTA, 119
National Aeronautics and Space Administration, 21, 177

National Environmental Policy Act (NEPA), 190, 208
national origin, 21, 187
National Research Council, 68
national security, 7, 19
natural gas, 5, 123
natural resources, 67
neutral, 138
New York Stock Exchange, 192
New Zealand, 110
Newsweek, viii, 97
nitrogen, 187
nitrous oxide, 188
North America, 33, 46, 66, 69, 119
North American Free Trade Agreement, 119

O

Obama, 31, 52, 61, 124
Obama Administration, 52
obstacles, vii, 29, 48, 129, 150
obstruction, 193, 208
oceans, 55
Office of Management and Budget, 20, 23, 24, 30, 51, 58, 59, 66, 71, 72, 82, 119, 123
officials, 22, 23, 26, 60, 64, 75, 76, 78, 79, 80, 81, 83, 84
oil, 88
openness, 21
operating costs, 21, 187
operations, 8, 12, 20, 25, 41, 46, 49, 57, 59, 67, 87, 88, 89, 100, 108, 119, 121, 132, 140, 150, 151, 164, 165, 197
opportunities, 3, 75, 76, 77, 82, 92, 94, 96, 98, 173, 180
Opportunities, 27, 82, 146, 160
optimization, 166
organic compounds, 14, 40, 72
organic matter, 185
organizing, 191
outreach, 108
overlap, 22
oversight, 21, 72, 101, 151, 164
ownership, 77, 104, 135, 151, 154, 155, 165, 168, 171, 176, 185, 187, 191
ownership structure, 151
oxidation, 185
oxygen, 41
ozone, 1, 4, 7, 10, 16, 17, 23, 24, 28, 30, 51, 52, 53, 54, 55, 70

P

packaging, viii, 13, 38, 66, 79, 94, 100, 102, 107, 108, 110
paints, 48
Pakistan, 78
parallel, 196, 198, 201
participants, 3, 21, 58, 70, 76, 79, 80, 82, 83, 84, 85, 94, 98, 105, 111, 138
penalties, 151, 190
permission, 211
permit, 8, 21, 43, 46, 60, 63, 65, 71, 78, 190, 194
petroleum, 5, 16, 72, 125
Petroleum, 33
Philadelphia, 141, 214
photographs, 203, 204, 205
pitch, 207
plants, 167, 185, 195
plastics, 79, 102
Poland, 110
polarity, 212
policies, viii, 21, 30, 44, 45, 46, 47, 48, 57, 59, 60, 61, 62, 65, 66, 71, 73, 82, 87, 89, 97, 98, 105, 132, 135, 136, 171, 187
policy, vii, 1, 2, 3, 4, 11, 17, 20, 21, 22, 23, 26, 29, 30, 31, 32, 34, 35, 44, 47, 48, 50, 52, 60, 61, 62, 63, 66, 68, 71, 73, 88, 104, 105, 109, 112, 136, 145, 155, 188, 194
policy initiative, 194
policy instruments, 52
policy issues, 32, 34, 44, 60
pollutants, 14, 20, 35, 40, 42, 43, 70
pollution, 3, 37, 40, 50, 52, 67, 91, 95
pools, 136, 173, 200
population, 70
portfolio, 119, 136, 184, 185, 192, 193
potential benefits, 43, 76
precedent, 84
preferential treatment, 8
preparation, 68, 143
present value, 119, 131, 158, 190
preservation, 137, 188
President, 2, 12, 22, 24, 63, 72, 80, 82, 99, 103, 104, 110, 124
President Obama, 2, 24, 63, 99, 110
prevention, 91
principles, 43, 50, 68, 91, 127, 134
private and public purchasers of electronics, viii, 109
private party, 187
private sector, viii, 30, 32, 91, 98, 101, 104
probability, 153
procurement, vii, viii, 1, 2, 3, 4, 5, 7, 8, 9, 13, 15, 16, 20, 21, 22, 23, 24, 25, 26, 29, 30, 31, 32, 34, 35, 36, 37, 39, 40, 44, 45, 47, 48, 49, 50, 51, 52, 55, 56, 57, 58, 59, 60, 61, 62, 63, 64, 65, 66, 68, 69, 71, 72, 73, 75, 76, 78, 79, 81, 83, 84, 85, 91, 98, 99, 100, 101, 104, 120, 121, 126, 129, 137, 161, 166, 167, 172, 180, 203
Procuring Solar Energy
  A Guide for Federal Facility Decision Makers, vi, viii, 117, 120
producers, 46
product performance, 43
productivity, 102, 103
profit, viii, ix, 45, 46, 108, 109
programming, 65
proliferation, 48
propane, 172, 218
property rights, 8, 192
proposed regulations, 62
prosperity, 125
protection, 2, 8, 20, 84, 193, 194
public access, 38
public discourse, 32
public health, 30
public housing, 93
public interest, 30
public policy, vii, 1, 3, 21, 105, 109, 194
public resources, 109
Puerto Rico, 24
pumps, 203
purchasing power, 63, 91, 107, 109
purity, 55
PVA, 119, 124

Q

qualifications, 26, 119, 151, 153, 166, 192
quality control, 46, 191
quality of life, 41

R

race, 8, 21, 25, 187
radiation, 41, 188, 193
radio, 57
rate of return, 119, 131
raw materials, 36, 37, 41
reading, 23
real estate, 176, 182, 187
real property, 158
reality, 41
recognition, 32, 57, 93, 94, 99, 114
recommendations, 25, 98, 99, 120, 126, 148, 176, 203, 208

recovery, 27, 102, 103
recyclers, viii, 78, 80, 84, 85, 109
recycling, 9, 27, 37, 39, 53, 57, 58, 70, 76, 77, 78, 79, 80, 81, 82, 84, 85, 95, 100, 102, 103, 107, 109
Reform, 21, 24, 72, 221
registries, 110
regulations, 2, 4, 8, 12, 15, 17, 22, 24, 26, 29, 34, 45, 48, 50, 55, 61, 62, 63, 71, 107, 123, 124, 187
reliability, 84, 125, 150, 164, 170, 176, 183, 193
relief, 27
remediation, 27, 200
renewable energy, 2, 103, 105, 119, 120, 122, 123, 124, 125, 126, 127, 132, 134, 135, 136, 137, 141, 143, 146, 154, 155, 156, 163, 164, 165, 166, 168, 170, 171, 173, 174, 175, 176, 178, 180, 183, 185, 187, 188, 191, 192, 194
Renewable Energy Optimization (REO), 126
renewable fuel, 16
replacement, 38, 134, 135, 140, 154, 155, 156, 160, 164, 173, 183
reputation, viii, 97
RES, 192
research and development, 37
researchers, viii, 109
resistance, 212
resources, vii, 3, 34, 37, 40, 42, 50, 57, 58, 64, 67, 69, 73, 75, 76, 77, 78, 79, 82, 84, 102, 103, 105, 109, 115, 124, 126, 130, 133, 139, 164, 172, 184, 185, 186, 188, 191, 192, 195
respiration, 185
response, 38, 75, 94, 114, 166
restrictions, 154, 155, 191, 194, 199
retail, 102, 192
retirement, 134
retrofitting, 67, 125
reusability, 70
rewards, 78, 99, 104
rights, 8, 132, 158, 192
risk, 21, 145, 147, 151, 152, 157, 161, 179, 198
risks, 16, 49, 186
rules, 66, 120, 134

S

safety, 23, 27, 43, 89, 149, 155, 163, 169, 175, 183, 185, 191, 199, 211
Samoa, 24
savings, 2, 18, 20, 24, 42, 46, 47, 51, 61, 65, 73, 76, 79, 80, 81, 83, 84, 87, 92, 95, 96, 99, 100, 101, 102, 103, 104, 105, 118, 121, 126, 127, 131, 164, 167, 169, 173, 175, 176, 187, 207, 216, 217
scaling, 204
scheduling, 191

school, 78, 79, 84, 88, 93
scope, 31, 43, 61, 71, 119, 130, 140, 142, 153, 157, 166, 173, 177, 188, 191
screening, 94, 120, 121, 126, 129, 130, 131, 132, 135, 138, 166, 203, 206, 207
Secretary of Commerce, 26
Secretary of Defense, 153
security, 5, 7, 19, 27, 96, 103, 105, 125
seller, 13
Senate, 20, 24
senses, 197
servers, viii, 57, 87, 94, 99, 105, 110
service life, 36, 45
service provider, 36, 52
services, 1, 2, 3, 4, 7, 8, 9, 13, 14, 15, 17, 18, 21, 22, 23, 24, 25, 26, 27, 28, 29, 30, 31, 32, 34, 35, 36, 37, 38, 39, 40, 41, 42, 44, 46, 47, 49, 50, 51, 52, 56, 57, 58, 59, 60, 62, 63, 64, 65, 66, 69, 70, 71, 73, 75, 77, 79, 85, 89, 92, 93, 96, 97, 98, 101, 102, 104, 114, 118, 120, 121, 138, 139, 141, 156, 164, 166, 167, 168, 170, 172, 173, 182, 187, 191, 195, 214, 219
sex, 8, 25
shade, 130, 131, 199, 210
shape, 2
shared multifunction printers, viii, 87
showing, 83
silver, 78, 83, 85, 111
Singapore, 110
small businesses, 1, 3, 4, 22, 69
smog, 67
SNS, 58
Social Security, 85, 112, 141, 214
Social Security Administration, 85, 112, 141, 214
software, 26, 69, 97, 103, 104, 105, 137
solar collectors, 194
solar space heating, 194
solar system, 123, 131, 132, 137, 142, 190, 196, 204, 208, 209
solid waste, 6, 47, 63, 80, 85, 94, 113, 122
solution, 55, 108, 109, 125, 135, 188
solvents, 54
Southeast Asia, 77
specialists, 173
species, 42, 67, 68, 188
specifications, 7, 9, 14, 23, 24, 25, 38, 47, 69, 94, 101, 113, 114, 115, 132, 142, 143, 145, 146, 148, 149, 156, 159, 162, 163, 169, 175, 183, 191, 210, 211
spending, vii, 1, 2, 29, 30, 51, 56, 58, 63, 65, 73, 75, 77
SSA, 112
stability, 49, 61, 62, 101

staff members, 135
stakeholders, viii, ix, 21, 36, 46, 99, 100, 102, 109, 110, 114, 115, 129
state, 2, 22, 25, 26, 30, 46, 47, 52, 64, 66, 67, 70, 73, 78, 79, 99, 103, 110, 131, 133, 135, 141, 142, 143, 152, 154, 165, 184, 186, 206
states, 24, 32, 38, 41, 66, 68, 69, 72, 73, 92, 108, 135, 150, 171, 184
statistics, 82
statutes, 4, 12, 13, 21, 22, 23, 24, 29, 31, 50, 51, 61, 62, 71, 72
statutory authority, 9, 62, 184
steel, 203, 212
storage, 38, 58, 105, 191, 194, 196, 197, 202, 203, 217
stormwater, 3
streams, 70
stressors, 67
structure, 131, 132, 136, 145, 155, 161, 178, 179, 184, 199
subscribers, 47
substitutions, 134
substrate, 197
success rate, 58
supplier, 27
suppliers, 4, 27, 184, 189
supply chain, 25
support services, 8, 23, 24
surfactants, 54
surplus, 34, 53, 78, 83
survey, 43, 47, 48, 187
sustainability, viii, 55, 65, 87, 92, 97, 98, 103, 109, 120, 123, 189
syndrome, 43

T

tanks, 203
target, 25, 34, 35, 59, 94, 127, 143, 156, 189
tariff, 132, 162, 171, 184, 188, 195, 218
tax incentive, 140, 151, 165, 187, 189, 191, 192
taxes, 185, 188
taxpayers, 125
team members, 129, 147, 161, 181
teams, 47, 193
technical assistance, 143
techniques, 37, 44, 68
technological advances, 38
technologies, 2, 3, 49, 73, 101, 102, 104, 105, 120, 186, 193, 194, 196, 197, 198, 200
technology, vii, 22, 29, 30, 35, 49, 56, 57, 58, 64, 75, 77, 79, 84, 92, 94, 97, 105, 120, 121, 125, 126, 129, 160, 184, 185, 189, 196, 198, 200, 203

temperature, 141, 184, 195, 198, 200, 213, 214
Tennessee Valley Authority, 112
Thailand, 110
The Green Electronics Council, viii, 47, 109, 110
thermal energy, 121, 123, 150, 151, 165, 194
thermodynamics, 192
thin films, 197
third dimension, 44
time constraints, 58
time frame, 7, 17, 19
total energy, 40, 82, 192
toxic metals, 94
toxic substances, vii, 41, 52, 71, 75
toxicity, 31, 47, 54, 78
tracks, 111, 125
trade, 31, 39, 43, 44, 45, 46, 70, 78, 92, 97
trade-off, 31, 43, 44, 46, 70
training, 15, 27, 28, 30, 47, 60, 88, 124, 138, 187
traits, 129
transaction costs, 126, 150, 152
transmission, 180, 186, 195, 219
transparency, 21, 91, 92
transport, 103, 186, 192
transportation, 25, 30, 38, 48, 57, 58, 60, 68, 73, 111, 198
Treasury, 56, 60, 80, 111
treatment, 5, 8, 187
tribal lands, 134
trucks, 42
TVA, 112

U

U.S. economy, 103
UK, 68, 69
UNESCO, 58, 68
uniform, 66
United, viii, 3, 12, 18, 23, 24, 25, 27, 33, 41, 47, 65, 66, 68, 69, 73, 75, 77, 80, 83, 91, 117, 118, 119, 123, 124, 130, 132, 136, 137, 141, 191, 221
United Nations, 65
United States, iv, vi, viii, 3, 12, 18, 23, 24, 25, 27, 33, 41, 47, 66, 68, 69, 73, 75, 77, 80, 83, 91, 117, 118, 119, 123, 124, 130, 132, 136, 137, 141, 191, 221
universities, 110, 193
updating, 84, 108, 109
up-front costs, 156, 185
urban, 198
US, viii, ix, 22, 91, 94, 107, 108, 109, 110, 112, 113
USDA, 17, 23, 24, 38, 43, 52, 53, 54, 55, 71
utility costs, 162, 187, 217
UV, 212

V

variables, 68, 200
variations, 32, 40, 146, 157, 159, 179
vegetation, 143, 207, 210
vehicles, 5, 9, 10, 13, 15, 16, 22, 28, 29, 38, 39, 51, 53, 59, 199
ventilation, 196, 200, 202, 207, 209
vessels, 203
Vice President, 97
video, 102
vision, 3, 125, 127
VOCs, 14, 40, 41
volatile organic compounds, 14, 40, 72
volatility, 127
vote, 25
voting, 147

W

wages, 21
warehouses, 202
warranties, 198
Washington, 21, 38, 67, 68, 70, 84, 105, 193
waste, vii, viii, 3, 6, 8, 23, 27, 29, 37, 38, 39, 41, 45, 47, 52, 55, 57, 63, 67, 70, 75, 76, 77, 78, 79, 80, 81, 83, 84, 85, 88, 94, 107, 109, 113, 200
waste disposal, 84
waste management, viii, 63, 75, 77
water, 3, 5, 14, 22, 24, 27, 32, 34, 37, 38, 39, 40, 42, 45, 46, 52, 55, 57, 61, 64, 67, 68, 69, 72, 73, 85, 113, 120, 123, 131, 141, 164, 165, 167, 170, 184, 185, 194, 195, 196, 199, 200, 201, 203, 214, 216, 217

water heater, 5, 22
water resources, 3, 40
water supplies, 42
waterways, 95
weakness, 198
wealth, 21, 25, 124
wealth distribution, 21
Web page, 14, 26
Websites, 26
welfare, 68
wetlands, 68
White House, 67, 68, 72, 85, 91, 96
wholesale, 192
wildlife, 180
windows, 32, 199
wires, 194, 196, 212
wood, 40, 43, 69
workers, 3, 39, 41, 197
workforce, 30, 64, 87
working conditions, 3, 21
working hours, 21
workload, 153
workstation, 88
World Trade Organization, 45
worldwide, 46, 113, 115

Y

yield, 5, 73, 103, 104, 105

Z

Zero Waste Alliance, ix, 109, 114